Anonymous

The Young Angler, Naturalist, and Pigeon and Rabbit Fancier

With hints as to the management of silkworms, the aquarium, etc.

Anonymous

The Young Angler, Naturalist, and Pigeon and Rabbit Fancier
With hints as to the management of silkworms, the aquarium, etc.

ISBN/EAN: 9783337034498

Printed in Europe, USA, Canada, Australia, Japan

Cover: Foto ©berggeist007 / pixelio.de

More available books at **www.hansebooks.com**

THE
YOUNG ANGLER

NATURALIST

AND

Pigeon and Rabbit Fancier

WITH

HINTS AS TO THE MANAGEMENT OF SILK-
WORMS, THE AQUARIUM, ETC.

With 98 Illustrations

LONDON
GEORGE ROUTLEDGE AND SONS
THE BROADWAY, LUDGATE
NEW YORK: 416 BROOME STREET

LONDON:
SAVILL, EDWARDS AND CO., PRINTERS, CHANDOS STREET,
COVENT GARDEN.

CONTENTS.

ANGLING:	PAGE
Fishing-rods	252
Fishing-lines	253
Floats	253
Reels	254
Hooks	254
Landing-net, Gaff, &c.	255
Common Baits and Ground Baits	255
Ground Bait	257
How a Fish is formed	258
The Stickleback, and how to catch him	259
The Minnow	259
The Bull-head	259
The Loach, or Groundling	259
The Bleak, or Blick	260
The Gudgeon	260
The Dace	261
The Roach	261
The Chub	261
The Carp	262
The Tench	262
The Perch	263
The Ruffe, or Pope	264
The Bream	264
The Flounder	265
The Pike	265
The Barbel	269
The Eel	269
Salt Water Angling	272
Natural Fly Fishing	272
Artificial Fly Fishing	274
Instructions for Artificial Fly making	274
Artificial Flies	275
Casting the Line	277
Monthly Guide for Bottom Fishing	278
Fishing Stations near London	279
Tributary Streams of the Thames	280
A List of some of the most celebrated Rivers of England, with the Fish which may be found in them	281

	PAGE
ANGLING (continued):	
Laws relative to Angling	282
Hints for Anglers	283
BRITISH SONG BIRDS:	
Bird-catching	285
The Breeding Cage	288
Linnets	289
The Blackcap	290
The Bullfinch	292
The Chaffinch	293
The Goldfinch	294
The Redpole	295
The Redstart	296
The Redbreast	297
The Skylark	299
The Titlark	300
The Woodlark	301
The Blackbird	302
The Thrush	303
The Canary	304
The Nightingale	306
Disorders of Caged Birds	308
ENGLISH TALKING BIRDS	310
PARROTS	313
DOMESTIC FOWLS:	
Common Fowls	315
Game Fowls	315
Dorking Fowls	315
The Bantam	315
Polish Fowls	316
Malay Fowls	316
Spanish Fowls	316
Persian Fowls	316
Cochin-China Fowls	316
How to choose Stock	317
How to feed Fowls	317
PIGEONS:	
The Stock Dove, or Wild Pigeon	320
The Turtle Dove	320
The Common Pigeon	321

CONTENTS.

PIGEONS (continued):
  The Fan-tail, or Broad-tailed
    Shaker 321
  The Narrow-tailed Shaker . . 321
  The Dutch Cropper 322
  The English Pouter, or Pouting
    Horseman 322
  The Parisian Pouter 323
  The Uploper 323
  The Horseman 324
  The Tumbler 324
  The Runt 325
  The Frill-back 326
  The Carrier 326
  The Mawmet 327
  The Barb 327
  The Spot 328
  The Dragoon 328
  The Jacobine 328
  The Capuchin 329
  The Ruff 329
  The Laugher 329
  The Trumpeter 329
  The Turbit 330
  The Owl 330
  The Nun 330
  The Helmet 331
  The Magpie 331
  The Lace 331
  The Finiken 331
  The Turner 331
  The Dove-Cote 331
  Feeding 334
  Mating 335

PIGEONS (continued):
  Diseases 335
  Laws relating to Pigeons . . . 336
  Conclusion 336

THE DOG:
  The Bloodhound 339
  The Wolf-Dog 338
  The Deerhound 338
  The Genuine Bulldog 338
  The Mastiff 338
  The Shepherd's and Drover's
    Dogs 338
  Pointers, Setters, Spaniels, Re-
    trievers, &c. 339
  The Terrier 339

THE SQUIRREL 340
THE DORMOUSE 342
WHITE MICE AND HARVEST MICE 344
THE GUINEA-PIG 346
THE HEDGEHOG 347

RABBITS:
  The Wild Rabbit 350
  The Tame Rabbit 351
  Fancy Rabbits 353
  Rabbit Hutches and Cotes . . 356
  Diseases 358
  Laws relating to Rabbits . . . 359

SILKWORMS 360
THE AQUARIUM 365
GARDENING 371

THE YOUNG ANGLER.

WE never met with a boy that did not like angling, and we believe that all boys have had a love for the gentle art, from the time when Simple Simon's mother, in the old rhyme, would not allow him to air his heels by a run to the river-side; so, being an ingenious youth, though not too bright, he filled the old woman's pail with water, got the poker for a fishing-rod, the clothes-line for a fishing-line, the meat-hook to hang his bait of fat bacon on, and like "Patience on a monument," sat bobbing for whale by the kitchen-fire.

But it is not in catching fish alone where the enjoyment of angling is found—it is so jolly to be out by the breezy river, to hear the birds singing beside the streams, to see the silver clouds, the blue sky, and the bending flowers—all mirrored in the bright, flashing waters, on which the gaily-painted float sits like a lily, rocked by the little eddies—to feel the warm sunshine and the gentle shower—these are the delights of angling, whether we catch many fish or few; for it is the holiday, after all, that is the great charm.

But there is no fishing without tackle, so we will begin with a brief statement of what is necessary for the equipment of the angler.

FISHING-RODS.

The ROD being the *staff* upon which the angler's sport in some measure depends, we shall give some particulars respecting its choice and manufacture. At all fishing-tackle shops, rods made of vine, bamboo, hazel, and hickory, of various lengths and fashions, may be procured; some are made to fit into canvas bags, whilst others resemble walking canes; the former, however, are decidedly the best, being longer and better made, as the joints are more carefully fitted together. The rod should, when put together, taper gradually from the butt end to the top, and be perfectly straight and even. For general purposes, a rod of about 12 feet in length is the most convenient; but in wide rivers, 15 and 18-feet rods are required. A bamboo rod with several tops of different degrees of strength, is well adapted for general purposes, and a cane rod is excellent for fine fishing. We must not, however, omit mentioning what is called "a general rod," which is so contrived, by means of top-joints of various degrees of length and elasticity, as to answer the various purposes of fly-fishing, trolling, or bottom fishing.

If the young angler wishes to turn rod manufacturer, he may use ash for the butts, and lancewood for the tops, and so make extremely good two-piece rods; or crab-tree for the stocks, with hazel or yew switches for the tops. A whalebone top is also an extremely good article; and should have a strong loop of horsehair whipt on it. Hazel wands are very serviceable additions to the stock of materials; they must be cut towards the end of the year, and allowed to dry and season in the chimney during winter, and if any accident should befal a good rod, a tolerable substitute may be made by sloping off the ends of three or four of these wands, and then fastening them firmly together with shoemaker's thread. It is a good plan to have a rod for each kind of fishing, as by such an arrangement they can be kept in complete order, and ready for immediate service. The rods should be ringed to guide the line from the reel; and when screwing the joints together, particular attention should be paid to these rings to see that they run regularly on the under side of the rod, so that there may not be the least likelihood of the line getting twisted. The rods should always be kept in a place of moderate temperature, neither too dry nor too moist; as in the former case they would become brittle, and in the latter, rotten; in warm, dry weather, if the joints are slightly shrunk, they may be moistened a little to make them adhere better; but if, through being too wet, they stick together so that you cannot readily take them to pieces, wait till they dry, rather than strain them by a forcible separation. It is a good plan to varnish the rods once in two or three years with copal varnish, or else with India-rubber dissolved over a slow fire in linseed oil; either of these preparations preserves the rods, but especial care must be taken, when re-varnishing, to scrape off the old surface before putting on the new; and the same precaution should be taken if the rods are carried to a fishing-tackle warehouse to be repaired. A single-handed fly-rod ought to be from 12 to 15 feet long, and as light and elastic as possible; a trout-rod for trolling with minnow, about the same length, but stronger; a rod for worm-fishing, the same; while a pike-rod ought to be strong, stiff, and as straight as a dart, and need never be more than 14 feet long; the rings through which the line passes ought also to be a good size and very strong, and the fewer of them there are on the rod the better. For roach and dace, the rod must be adapted to the fishing-ground, for sometimes the angler is stationed on a bank, behind a foreground of reeds, flags, or willows, and his rod must be long enough to reach far beyond these obstacles; so that 20 feet is not a bit too long for these wild, sedgy embankments, under which roach and dace delight to shelter. The best rods are those made of ash and lancewood.

FISHING-LINES.

The most serviceable lines are made of pure horsehair, for such as are composed of hair and silk, from retaining the water, soon become rotten; neither can they be thrown with the same precision, as they get soft and flabby, and fall heavily on the water. Good lines should be perfectly twisted, round, and without any irregularities, and in point of colour those which are of a light grey, or brown, or white, are the most useful; some anglers, however, prefer a light sorrel tint. The bottom, or casting-line for fly-fishing, which i affixed to the line on the reel, must be of gut, and of about th same length as the rod; the gut should be strong at the top, anw very fine at the "dropper," or bottom, and before any flies are made upon it, it should be picked and tried to see that it is of an uniform thickness throughout.

It is hardly worth a lad's while to attempt manufacturing fishing-lines, as they may always be purchased more neatly fabricated, and at a much cheaper rate than he can make them. When fastening the line on the rod, the loop of the line should be passed through the ring at the end of the top joint, carried over the ferrule, and then drawn up to the top again, by which plan the loop will be secured, and the line hung from the extreme ring.

Blakey, a great authority, in his deservedly popular work on angling, says he prefers the old "cast-line of about 4 or 5 feet in length, and from 4 to 6 or 8 hairs in thickness, on which to place the gut and flies, as a line so prepared can be thrown much truer to any given point." Lines for trolling are not required to be so light and elastic as those used in fly-fishing. Horsehair lines are best.

FLOATS.

Floats can always be procured ready-made, of all sizes and every variety of shape. For small fish and slow streams, quill floats will be found the best; and in strong and rapid rivers, or for the larger kinds of fish, cork floats must be employed. If the young angler prefers making cork floats to purchasing them, he must procure a piece of fine-grained sound cork, and bore a hole through it with a small red-hot iron, then put in a quill which will exactly fit the aperture, and afterwards cut the cork into the shape of a pear. When this is finished, he must grind it smooth with pumice-stone, and paint and varnish it; and if he uses two or three bright colours in the painting he will add much to the gaiety of its appearance. The cork float should swim perpendicularly in the water, so that it may betray the slightest nibble, and must be carefully poised by fastening a few shot on the line; the sizes of shot proper for this purpose are from swan shot down to No. 4; they should be split about half-way through with a small chisel, so as to make a gap sufficiently wide to admit the line, and when the latter is put in, the gap should be closed with a pair of pliers, though the teeth will do.

REELS.

A reel is very useful, as with its assistance parts of a river may be reached which could not otherwise be attempted, it enables the angler also to play his fish with the greatest ease and certainty. When purchasing a reel, a multiplying one should be selected, as it is superior to all others, and enables the angler to lengthen or shorten his line rapidly. It must be kept clean and well oiled, and great care taken that no grit of any kind gets into it.

HOOKS.

Hooks are of various patterns and sizes, beginning at No. 1, which is the largest salmon size, and ending at No. 14, called the smallest midge. The round-bend hook is the shape most used in England, while in Scotland the sneck-bend appears to be the favourite. Limerick hooks are excellent; and those made in Dublin, marked with 2 F's, 2 B's, and so on, are second to none. A bad hook, be it remembered, is worse than a bad knife, only fit to be thrown away. The following table shows the sizes of the hooks most suitable to the various fish:

Barbel, 7, 8, 9.	Flounders, 3.	Perch, 7.
Bleak, 11, 12, 13.	Grayling, 10.	Roach, 10, 11, 12.
Bream, 10.	Gudgeon, 9, 10.	Rudd, 10.
Carp, 7, 8, 9.	Loaches, 13.	Ruffe, 10.
Chub, 8, 9.	Miller's thumb, 13.	Smelt, 9.
Dace, 10, 11, 12.	Minnow, 13.	Tench, 9, 10.
Eels, 8.	Midge, 14.	Trout, 6.

When fastening the hooks on your lines, use strong, but fine silk, and if you can get it near the colour of your bait, so much the better; wax the silk thoroughly with shoemaker's wax, and wrap it four or five times round the body of the hook, then place the gut or hair on the inside of your hook, and continue winding the silk tightly round till you have wrapped it about three parts down the hook.

Whipping is finished off by slipping the end of the silk through the last circle, and drawing it tight. Knotting, by laying two pieces of gut or hair together, one overlapping the other some three inches or so, then holding one end in the left hand, while forming a simple slip knot on it; then turning the other end to the right, and doing the same; after that drawing the two together, which makes the knot complete. No direct pull will ever unloosen this water-knot, though it can be undone easily. Gut is obtained from the silkworm. Gimp is any kind of tackle covered with fine brass wire, to protect it from the teeth of fish, sharp stones, or other injury.

LANDING-NET, GAFF, ETC.

The landing-net is simply a hoop with a handle to it, to which a net is fixed, to lift out the fish, when hooked, without loosing it, or breaking the tackle. The gaff and landing-hook are used for the same purpose. The basket or creel, as every boy knows, is slung over the shoulder with a belt, and may be bought big enough to hold a whole family of salmon, or little enough to hold all a boy at first catches, and which he might put in his eye and see none the worse for. Bait kettles are made of tin, and if you haven't one, go into the kitchen, and take the cook's flour dredger, it is a capital make-shift. Drag-hook, clearing-ring, and disgorger must be seen to be understood; any clever boy used to angling will show you how to use them. The drag-hook will pull up a weed in which your fishing-hook has got entangled. The clearing-hook is for a similar purpose, and often saves the casting line when the hook is fast.

COMMON BAITS, AND GROUND BAITS.

Fish, in their natural element, take such baits as the changing seasons produce, and will not at one time of the year bite at the same bait which they may be caught with at another; for instance, in spring and autumn, worms may be used all day long, and night too, if you can keep awake long enough to fish while the moon and stars are shining; but in summer, worms must only be used early and late, morning and evening. An earth-worm is naturally the first bait the young angler looks out for; it is always to be had, is put on the hook without difficulty, and (excepting at the times above stated) may always be used for certain kinds of fish, with the certainty of hooking something, if proper patience is used. The dew-worm or garden-worm every boy knows, also the marsh-worm or blue-head, which is found at night, by taking a candle and lantern, in moist, undrained places; while the tag-tail must be sought for in strong clays, where turnips and mangold-wurzel are grown; and the brandling in any kind of decaying vegetable matter; as for the red-worm, that is always to be dug out of sewers and the banks of ditches. When baiting with a worm, the hook should be put in close to the top of the worm's head, and then passed carefully down, gently working the worm up the hook at the same time. Not more than a quarter of an inch of the worm should be left hanging over the hook. To scour or starve these worms, and get rid of the earthy matter they contain, they must be placed in damp moss, not soddened with water, remember, but only damp. In creeping through the fibres of the moss, they compress and empty themselves.

The ash-grub, which is found in the rotten bark of a tree that has been felled some time, is an excellent bait for grayling, chub, dace, or roach, and may be used all the year round; it should be kept in wheat bran.

The brandling worm is a capital bait for almost any kind of fish.

The name of the cabbage worm indicates its habitats; it is a good bait for chub, dace, roach, or trout.

The caterpillar, which may be found in the leaves of cabbages, is employed for the same fish as the cabbage worm.

Of the cod bait, or caddis worm, there are three kinds; they may be found by stony brooks, pits, or ponds, and in ditches. They are good for trout, dace, chub, bream, bleak, roach, and grayling.

The cow-dung bob is found under cow-dung, and is somewhat like a gentle in shape, but larger; it should be kept in earth. Chub, carp, tench, roach, dace, and trout will take this worm eagerly.

The locality of the crab-tree worm is indicated by its name; it is a good bait for roach, dace, trout, and chub.

Flag or dock worms inhabit the fibres of flag roots in old pits or ponds; they are excellent baits, and may be kept in bran.

Every boy knows how gentles, or maggots, are bred. A little bran and damp sand must be put in the vessel in which they are kept, for the purpose of scouring them; they are tempting baits for all kinds of fish. When putting a gentle on the hook, you must insert the hook at one end of it, and bring it out at the other, and then draw the gentle back until it completely covers the point of the hook.

The garden worm, to be good, should have a red head, a streak down the back, and a broad tail; it makes a good bait for chub, eels, perch, or barbel.

The marsh worm is a good bait for trout, perch, grayling, or bream, but it must be scoured for a longer time than the brandling.

Oak worms may be gathered on the leaves of the oak tree, and are good baits for chub, dace, roach, or trout.

Palmer worms, or cankers, found on herbs, plants, and trees, are excellent baits.

The tag-tail is accounted a good bait for trout in cloudy weather, or when the water is muddy.

White grubs or white bait are much larger than gentles, and may be found in sandy and meadow land; they are good baits for chub, roach, bream, tench, trout, carp, and dace; and should be kept closely covered in an earthen pot with the earth about them.

Wasp-grubs may be taken from the nest; they require to be hardened for half-an-hour in a warm oven, and are good bait.

House crickets are good to dib with for chub.

Beetles are good also for chub; they may be found in cow-dung.

Miller's thumbs, bleaks, minnows, dace, gudgeons, loaches, sticklebacks, smelts, and roach, are used as baits for larger fish.

Grasshoppers are good baits during June, July, and August, for roach, grayling, chub, and trout; their legs and wings must be taken off before they are put on the hook.

Salmon spawn is an excellent bait for trout and chub, and may be purchased at the shops ready for use.

Cheese Paste.—Take some old Cheshire cheese and the crumb of white bread, and mix them up to a tolerable degree of consistency, and you will make a good bait for chub.

White Bread Paste.—Knead crumbs of white bread dipped in honey in the palm of your hand until they attain a fair degree of consistency; it is good for tench, carp, roach, and dace.

Wheat Paste.—Procure some new wheat, remove the husks, and afterwards pound it; then pour some milk or water over, and gently simmer the composition; when cold, it will be somewhat like a jelly, and a very small piece only should be put on the hook.

Sheep's blood and saffron make a good paste for roach, bleak, dace, perch, and trout.

For barbel, an excellent paste may be made by dipping the crumb of new white bread in the liquor in which chandlers' greaves has been boiled, adding a little of the greaves, and working it up till stiff.

Paste baits are not at all adapted for swift, running streams, but for quiet brooks, ponds, or very still rivers; you must be quick of eye, and sharp to strike, otherwise both fish and bait will give you the slip. A quill float is better than a cork one when baiting with paste, as it betrays the slightest nibble.

GROUND BAIT.

Ground baiting is a most essential part of angling, and ought never to be omitted, as success in bottom or float fishing cannot be expected, unless the proper means for drawing the fish together are resorted to. The object for throwing bait into the water, is to collect fish to one particular spot, and then to use a superior kind of bait, though of a similar kind, on the hook. Thus, if going to angle with earth worms, throw in for ground bait those that are unscoured, and fish with those that are well scoured.

For barbel, it is necessary to make the lumps of ground bait large in proportion to the strength of the current in which you fish; chop or break a pound of greaves into small pieces, and pour hot water over it, let it remain till it softens, strain the water away, and work it up with clay into lumps or balls, and add a little bran to it.

For chub, roach, and carp, mix bran and clay together into lumps about the size of an apple; place some gentles in the middle, and close the clay over them. It is a very useful bait in a still pond, hole, or slight eddy.

For roach, dace, and bleak, work some clay and bran together into balls, about the size of a pigeon's egg.

For chub, carp, roach, and dace, take the crumb of white bread, soak it in water, squeeze it almost dry, add bran and pollard, and work them up together until they acquire the consistency of clay.

For carp, tench, eels, perch, and bream, fresh grains will be found very serviceable; they must be perfectly fresh.

Gentles and worms may be thrown in without taking the trouble of working them into balls or clay, if the water is perfectly still; but if you are fishing in a stream, such a system of ground baiting is injurious, as the gentles are carried away by the stream, and draw the fish from the spot.

HOW A FISH IS FORMED.

Before giving our young readers a description of the different varieties of fishes they may chance to take at times, we shall first attempt to show what a fish is like, and why it is neither flesh nor fowl, though some say the sturgeon partakes of all three. Most fishes are covered with scales; as the knights of old were sheathed in scale armour, so are fishes protected by these scales. They have a backbone, and instead of lungs breathe through their gills, through which pass the water and air they take into their mouths; and that is the way they live and breathe. They have also a "swimming bladder" under the spine, by compressing or expanding which they are enabled to sink down like a leaden bullet, or rise up light as a bird springing into the air. Their very forms are built for swimming, and their motions in the water in some measure resemble that of birds in the air; their fins are their wings, their tails the rudder, and they use the latter as a boy does a single oar in the stern of a boat, and can row themselves along with it at immeasurable speed. You all know the prickly fin on the back of the perch—that is the dorsal fin of a fish; that before the little one next the tail is the ventral fin; and these balance the fish, and thus prevent it rolling over; while the pectoral fin, the one on the breast, is used by the fish in pushing itself forward, and is a kind of screw in the fish engine, while the caudal or tail fins are paddles, sails, screws, rudders, oars, or anything you please to imagine, that gives the fish both steerage and rapid motion. The eye of the fish is beautifully adapted for seeing in the water.

Every boy, though he has eaten only a red herring, knows what the roe of a fish is. The codfish and salmon contain millions of these little eggs, and were they all to come to life, and neither be devoured nor destroyed, instead of not being able to see the wood for trees, we should neither see the river nor the ocean for fishes.

We shall now proceed to give an account of the different kinds of fish which are found in the rivers of Great Britain, and the best means to be adopted to catch them; and we will finish each fish as we go on, leaving only the bones.

THE STICKLEBACK, AND HOW TO CATCH HIM.

This little quarrelsome fellow—for he is a terrible fighter, and would kick up a shindy were he as big as a whale—is seldom more than two inches long, and marked with the most beautiful crimson, green, and golden colours the eye ever delighted to dwell upon; but only put your stick in the water, and he will fight that if he can find nothing else to battle with, and fetch it a famous knock, too, with his little hard head. If once he is heartily thrashed by a rival stickleback, he hides his diminished head in any hole or corner, loses all his gaudy colouring, and, like a threadbare quaker, comes out again in a suit of seedy-looking dull grey. When fishing for these small fry—in which we include minnows—rods, lines, and hooks, should all be the finest and most delicate that can be purchased, and the hooks especially the very smallest that are used. It is a good plan to whip three or four hooks on fine gut, or strong horsehair, using a short line of horsehair or silk, letting the hooks hang some three or four inches below each other, for when they bite well three or four of these little hungry fishes may be pulled out at a time. A crowquill float is quite big enough, and as for bait nothing can be better than little pieces of the small red worms.

THE MINNOW

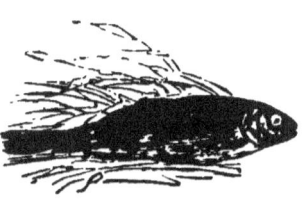

Is sometimes called the "pink," also the "mennow," and mostly found in clean swift brooks; is very partial to company, for where there is one, a hundred are not far off. In the gloomy days of winter it hides itself under the mud, or among weeds. As for colour, it is sometimes met with of beautiful pearly-white, at other times and places blue and green on the back, and red or white on the belly, and sometimes it is found with a tinge of yellow. The pearly-white is preferred when it is used as a bait for trout, salmon, or pike. May be fished for with the same tackle and bait as the stickleback.

THE BULL-HEAD,

So called on account of its great head; also "miller's thumb," because its head is flat and wide, as if one of its bullheaded ancestors had been taken out of the mill-dam, ages ago, and pressed under the heavy thumb of a miller, then let go to increase and multiply the race of flat-heads. He is very partial to shoving his ugly head under a stone, and leaving his tail out, which a quick-handed boy very often seizes, and drags him out. As it is so fond of the bottom, the bait must only just clear the ground, and a small piece of worm is as good a bait as can be used.

THE LOACH, OR GROUNDLING,

Is about the same size as the minnow, of a dusky brown colour, with a compressed head, and a beard formed of six fleshy tufts that hang from its lips, and in some places, through this pecufiarity, is called "beardie." For its size it has an immense breadth of tail adjoining the spine, and can never be mistaken for any other fish. The word groundling sufficiently indicates its habits, telling that it must be sought for at the bottom, almost in the same way and with the same bait as the bull-head. It is sometimes used as a bait on night-lines for large eels, and that is all we ever heard it was good for.

THE BLEAK, OR BLICK.

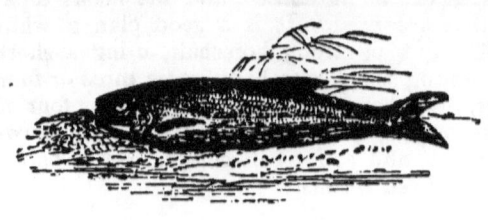

This clean, beautiful, lively little fish — the favourite of every young angler — is seldom more than four or five inches long, and is constantly to be seen in large shoals near the surface of almost every still-flowing river. The head is small and pretty, the eyes prominent, with a ruby-coloured patch below, while the back is a beautiful olive-green, and the sides and belly of a silvery whiteness. For its size the scales are rather large, the fins transparent, and the tail forked. It is pleasant to watch the movements of a shoal of bleak in clear water, where they may be seen swimming round and nibbling at the baits, for in bleak fishing half-a-dozen hooks may be used at a time, as in minnow fishing, with gentles, red-worms, caddis-paste, &c., for bait. In cold weather they swim deep. We know no better practice for the young fly-fisher than to whip for bleak in warm weather, when they swim near the surface. A small black gnat is the best for this purpose.

THE GUDGEON

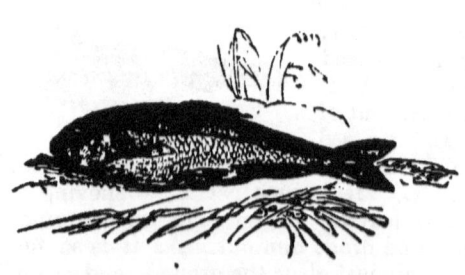

Is another of the small fry, seldom exceeding five or six inches in length; has a large thick head, a round body, and a beard on the upper lip. The colour of its back is a pale brown, of its belly a reddish white, while its fins have an orange or reddish yellow tinge, and both the dorsal fin and tail are spotted with black. Gudgeons are fond of swimming together in shoals at the bottom of gravelly brooks and rapid rivers. They are very rarely seen on the surface. Though little, it is an excellently flavoured fish, and a hungry boy would easily eat up a round score at a meal if nicely cooked. In fishing for gudgeon the tackle should be as fine as that used for the minnow; but the hooks must be No. 8 or 9, while the most killing bait is the red-worm, next to this the gentle, then the caddis-worm; if these are not to be had, almost any kind of paste will do. What would scare away any other fish draws the gudgeon to the spot, and that is stirring up the bottom of the brook or river with a ground-rake. It is better than all ground-baiting.

THE DACE

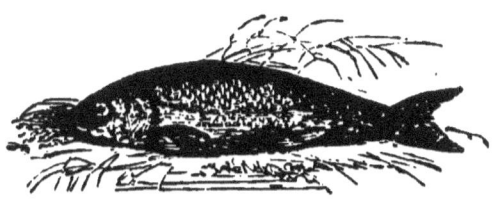

Is a beautifully-shaped fish, and is rarely found more than nine or ten inches long, mostly in shoals, like birds of a feather that flock together. The tail is forked, the body slender, the head small. In colour it is dusky on the back, varied with olive-green patches, with sides and belly of a bright silver hue, fins red, but not so deeply dyed as those of the roach. It prefers swift, gravelly streams in which there are plenty of weeds, and is found everywhere. The dace is a very strong fish for its size, and when hooked, struggles as hard as the trout to escape. Both dace and roach are fished for in the same way, and with the same tackle. A light stiff rod seventeen or eighteen feet long, so that the angler may stand so far back as not to show himself, and yet be able to drop his bait gently in the water. A line very fine down to the swan-quill float, below the float single hair or fine gut, the knots unwhipped to be invisible. No. 10 or 12 hook, baited with gentles, paste coloured with vermilion, worms well scoured, grubs or salmon-roe.

THE ROACH.

The roach, as every boy knows, has a funny-looking, round, leathery mouth, and may be a good whistler, though we never heard him; his teeth are placed in his throat, so that he has the pleasure of first tickling his palate, then enjoying his food in his throat, and bolting it afterwards at his leisure. His beautiful large scales have a pale golden tinge, which almost deepen to brown on the back, while his fins and iris are red as a summer rose, or richly stained verbena. He is a fine, deep fish, and is sometimes found to weigh as much as two pounds, and more than that, it is said, in the rivers on the Continent. In fishing for the dace we have described all that is necessary to be known in roach-fishing.

THE CHUB,

Though a strong fish, is very timid, and retreats into the deepest hole it can find, when apprehensive of danger. It is short, thick, and high-backed, has large scales, has a greenish-brown back and

head, with silvery-brown sides that almost approach to yellow as summer advances, silver belly, yellow breast, fins, and a brown forked tail. It is generally caught from one to four pounds weight, seldom beyond the latter. Young chub are often mistaken for dace. It is a famous fellow for breaking lines, and if it has a chance, will run with the hook under weeds and old stumps, and pull like a cart-horse. It is a bold biting fish, and in summer bites during the whole of the day, but best in the morning and evening; it may also be taken in the night time. It is very bony, and not good food, but there are people who say he is eatable, and you all know that there is no accounting for taste. Look down our list of baits for those adapted for chub. Throw in plenty of ground-bait, made of soaked bread, bran, and pollard, well worked together.

THE CARP,

Like the roach, has his teeth—if the bony apparatus may be so called—in his throat. His back—which is of a dusky-yellow colour, almost approaching to brown—is arched and thick, while his belly is white; he has also a shortish beard on each side of his mouth. He appears to be rather a dainty gentleman, taking a bait at one time which no enticement will tempt him to taste at another. Left to himself, however, he generally feeds on worms and insects, brings up a large family, and if not caught, lives to a good old age. It is a very shy, cunning fish, and from its extreme craftiness has been styled the water fox; is found in lakes, ponds, and rivers, and frequents the quietest and deepest parts of the stream, especially holes near flood-gates, and beds of weeds. The best time to angle for this fish is either very early or very late, as it seldom bites in the middle of the day, unless a shower happens to fall. Use a long, light rod, with a reel, and let the line be of the finest description; the hooks, if worms are employed as bait, should be No. 5 or 6; if maggots, No. 8 or 9; and if wasp-grubs, No. 7. Refer to the list of baits at page 5.

THE TENCH

Thrives best in stagnant or slow-changing ponds that have a rich loamy soil. It is a thick fish, and greatly resembles the carp in shape, and, like the carp, has no teeth in its mouth, but a similar bony formation in the throat; it has also a small thin beard at each corner of its mouth. In colour its body is a deep olive tinged with gold, the gill-covers bright yellow, fins a dark brownish purple, and the tail square, while his scales are thin and covered with slime, telling how fond he is of  burying himself in the mud. Like the eel, he buries himself in the mud in winter, but during the hot days of summer or autumn you are pretty sure to find him near the surface. Like the carp, he will live some time out of water, and may be carried a long way if wrapped up in wet grass. From the rapidity with which the tench darts off if a noise is made near him, it would not seem unreasonable to conclude that he can hear, though sound may give some peculiar vibration to water which we know nothing of. When hooked he will attempt to rush into the mud, so that a firm hold must be kept to keep his head up and his mouth open without straining the line too hard. He is fished for in the same way as carp.

THE PERCH

Is a fine, formidable, handsome-looking fish, with his armed dorsal fin sticking up with its line of spines sharp as the points of great steel needles. His high-arched back is of a deep olive-green colour, from which descends broad black  bars, which gradually fade into white towards the belly; the underneath fins nearest his tail are of a rich scarlet, while the tail itself is hardly of so deep a colour. He is a dangerous fellow for a novice to lay hold of, and many an "Oh, my hand!" has his sharp spines caused to be uttered, while the young angler went dancing about in agony, requiring no fiddler to play to him. He thrives best in tidal rivers where the turn of the water is saltish, and is taken there much larger than in wholly fresh rivers. The perch is many years arriving at its full size, and in point of flavour is surpassed by none of the fresh-water tribes. It takes a bait very freely. Strong tackle is necessary in angling for it, gut or twisted hair-line, cork float, and No. 7 hook. Perch lurk near bridges, mill-pools, and locks, in navigable rivers and canals, and in other streams, near rushes, in dark still holes and eddies, and in the gravelly parts of rivers. Dark, windy days, if the weather is not too cold, are as good as any for perch fishing.

THE RUFFE, OR POPE,

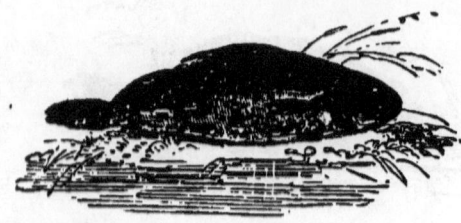

Is by many considered as only another variety of perch, and certainly bears a great resemblance to it, having also a high dorsal fin armed with sharp spines; but its colour is different, being a dusky olive varied with black spots; and most naturalists have classed it in a different subdivision of British fishes. It abounds most in slow, deep, quiet rivers that have a loamy bottom, and must be fished for in those still, deep places which are so often found near jetties, or where the banks have been washed away year after year, and holes deep as wells have been scooped out by the action of the water, but which are now quiet and calm as an undisturbed rain-waterbutt. In angling for it, use a quill float and a No. 7 hook; the moment you observe a bite, strike, without allowing much line. The proper baits are small red worms and brandlings, and they should be suffered to drag lightly on the ground; throw in a ground-bait made of clay and worms, if the water is clear, but if it is muddy, worms alone will do. This fish will bite freely at any time of the day during summer, but mostly in cloudy, sultry weather.

THE BREAM

Is more like a flat-fish than any other frequenting our English rivers, excepting the flounder. One of our popular writers on angling says he is like a pair of bellows in shape, and much of the same flavour. The bream is very narrow across the back, which, as well as the belly, is greatly arched—so much so, indeed, as to make the form of the fish nearly an oval. He has a little bead-pointed snout, small mouth, and no teeth, while in colour his back is of a bluish iron-grey, inclining to white on the belly. Like the pike, he loves to frequent still water, and though he is seldom taken above one pound or two in weight, yet he has been known to grow to an enormous size. The bream is principally found in large lakes and still rivers. It is best to angle for it in May, when it is in its prime, and from the end of July to the end of September; and in these months from sunrise till eight o'clock in the morning, and from five o'clock till dusk in the evening. Use a gut line, quill float, and No. 10 hook, and let the bait touch the bottom, or nearly so. The angler should be very silent, keep from the edge of the water as much as possible, and strike the instant the float is drawn under the surface of the water.

THE FLOUNDER,

Though a sea-loving fish, is found in some of our English rivers, especially the Thames, where it affords great sport to the London boys. As everybody knows, it is a flat fish, the upper part of a dirty-brown colour dashed with dusky yellow spots; the belly is white, while round its body there runs a row of sharp spines, which distinguish it from all other kinds of flat-fish. In weight it is seldom caught above 2 pounds, though it grows much larger. It may be taken from March to August, but as their spawning time is in June, they should not then be eaten. Small red worms, marsh worms, and brandlings, are the best baits, and they should be put on No. 6 hooks. Let the bait touch the bottom, and keep it continually moving, as these fish are exceedingly cunning.

The SMELT abounds in so few places, and the Rudd is a fish so little cared for, that here the boy's first lesson in angling may be considered to end, as none of the fishes we have hitherto described are either very large or difficult to capture, and are generally as commonly to be found "as way to parish church." We now come to sharp-biting Pike, great, bony Barbel, and slippery Eels. Pike, Barbel, and Eels are common enough, but we have declined entering them under the head of fishes already given, as they require different tackle and different management; and as a boy unused to angling would not only be likely enough to lose his line every time he hooked a fish, to say nothing of the top of his rod, and perhaps himself. We will suppose him to have become proficient through fishing for the "smaller fry" already described, before commencing with the fresh-water giants he is now about to be introduced to. As for Salmon, Trout, &c., they belong to too high an art of angling to have a place here.

THE PIKE

Is found in ponds, brooks that empty themselves into rivers, and also in rivers. It is called a jack until it weighs 4 pounds; when it has attained that size, it becomes a pike, "a fresh-water shark," a devourer of almost every fish that it can get into its mouth, and a terrible mouth the pike has, I can tell you; for its long, frightful jaws are armed with several hundreds of sharp teeth, and it can bite like a shark too. It has an ugly, savage-looking head, though it is rather prettily marked, being of a pale olive-grey colour, which deepens on the back, while it is dashed on the sides with yellow spots of all forms. The size it reaches at times is enormous, having frequently been caught a yard long.

Pike are in season from May to February, and the best time to take them with the hook is early in spring, before the water-weeds begin to shoot, and also in October, after the weeds have begun to rot, for he is in fine season at this period—firm and fleshy, and as voracious as a starved wolf. As for baits, we hardly know what it will not take, as all are fish that come to its net. The baits generally used in fishing for it are roach, dace, gudgeon, minnows, chub, bleak, young frogs, lob-worms, fish or flies, real or artificial, and the proper size of a bait is when it weighs from one to four ounces. There are several methods of trolling for this fish, namely,

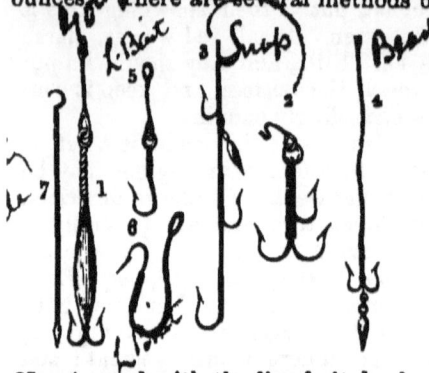

with the gorge hook, No. 1 in the annexed illustration, which is loaded on the shank with lead; with the snap hook, either spring or plain, composed of three hooks fastened together, Nos. 2, 3; with the bead hook, formed of two single hooks, tied back to back, with a little drop or bead of lead affixed to a link or two of chain depending from the lower part of it, No. 4; and with the live-bait hooks, which may be either single or double, Nos. 5 and 6. In baiting these various hooks, the following directions must be carefully attended to:—

The GORGE-HOOK. Hook the curved end of a baiting-needle, No. 7, to the loop of the gimp on which the hook is fastened, pass the needle through the mouth of the bait, and bring it out at the tail; the lead on the hook will thus be hidden in its belly, and the barbs or shanks inside its mouth; and in order to keep the bait steady on the hook, it is a good plan to tie its tail to the gimp with some white thread.

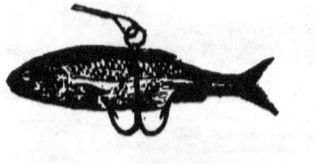

The SNAP-HOOK is baited by thrusting the point of the upper or small hook under the skin of the bait, on the side, and bringing it up to the back fin. Another snap-hook is baited by passing the loop of the gimp inside the gill of the bait, and bringing it out at the mouth; the lead thus lies in its throat, the first hook outside its gill, and the others in its side, the barbs being just beneath the skin; the bait's mouth should next be sewn up, so as to keep the lead and hooks in their proper places.

On a BEAD-HOOK, a gudgeon or barbel is the best bait; the little drop or bead of lead should be put into its mouth, which should afterwards be sewn up with white thread.

The LIVE BAIT must have a No. 3 or 4 hook passed either through its lips or the flesh beneath the back fin ; in the latter  plan, care must be taken not to touch the back-bone, or the bait will soon die. If the live-bait seeks the weeds, it must be stopped, and should it become sluggish, a good shake of the rod will stir it up. It is best to use a float with live-bait, as the length of line renders it difficult to heave the bait in mid-water at a proper depth without.

The rod for trolling must be very strong and stiff, about 14 or 16 feet in length, and have a whalebone or hickory top ; the line must be at least 30 yards in length (sometimes it is full 60), made either of silk, or silk and gut twisted together, and be kept on a winch. When you begin trolling, first fasten the winch on to the rod, then pass the line through the rings on the under side of the rod, and attach the hook to the line by a small swivel ; next grasp the rod in your right hand, just above the winch, and rest the butt-end of it against the side of your stomach, draw out, with your left hand, a yard or two of the line from the swivel, hold it firmly, and then with a sharp jerk from your right hand, cast the bait into the stream, and let the line which you hold in your left hand run out freely, that the hook may not be checked when cast out, by your holding the line too fast, and so fall short of the spot you wished to reach. Let the bait touch the water very lightly, and allow it to sink nearly to the bottom; then draw it gradually to the surface, and continue moving it in this manner till you feel a bite ; you must then let out your line quickly, give the fish about five minutes to gorge, strike, and draw in until you see your prize ; play him very carefully, and keep him away from weeds or piles, or anything likely to endanger the safety of your tackle. The pike requires some time to bite fairly, as he is apt to blow the bait out of his mouth ; that is why he must have so much time allowed him to gorge the bait. Many a young angler loses his fish through beginning to draw in too soon.

When you fish with a live bait, put a gudgeon on the hook, it being a strong fish, and one that will live some time ; use a middling-sized barrel-shaped cork float, and put a few swan-shot on the line, and adjust the float so that the bait may hang about mid-water ; let it float about for a few minutes, without taking it out, unless it gets amongst weeds or too near shore, and the instant the pike seizes the bait, which he does with much violence, let your line free, give him a few minutes to gorge, and then strike. When fishing with a snap-hook, either spring or plain, allow the fish no time to pouch, but strike immediately you feel a bite. If you troll with the bead-hook, throw it in as directed for the gorge-hook, draw it frequently to the surface, and let it sink gradually again. You may now and then take it out of the water, and cast it into a fresh place, and so fish every yard of the stream where it is probable a pike may happen to be; when you feel a bite, let the fish run, and give him time to gorge, before you strike.

The favourite haunts of the pike are the deep eddies in tumbling-bays, and deep still water in rivers; near beds of candock weeds, and mouths of ditches or small streams which empty themselves into rivers, and near flood-gates, and close to beds of bulrushes in lakes and canals. When the water is muddy in rivers, the pike makes its way into the little streamlets, as the water in those places is tolerably clear. Pike feed at all times of the day, but bite most freely during a breeze; in stormy, chilly weather you may troll for them, when all other fish refuse the most enticing baits; if frosty, or when northerly and easterly winds set in, you must not calculate on much sport; but directly the wind shifts to the south, pike bite readily. When you use live baits, take at least six in your kettle, and give them fresh water often; if you intend to employ gorge-hooks, bait three of them before you begin, and keep them in bran in a bait-box, large enough for the baits to lie at their length; always have fresh and lively baits, for the pike is extremely fastidious in his taste.

THE BARBEL

Is a large, well-formed, powerful fish, and if you want to try the strength of your tackle, hook one about seven or eight pounds weight, and he'll find you work enough for both elbow and wrist. He takes his name from his beard, which stands out pretty prominently from each side of his upper lip. His general colour is a silvery grey, which darkens on the back, and is nearly white on the belly. The scales are rounded; the dorsal fin small, and of a bluish brown; his other fins are also brown, but finished off at the tips with yellow; while his tail, which is forked, is of a purplish brown. He likes to feel the water stir about him, and is very partial to rapid rivers which roll over stony bottoms; old walls, sunken piles, and shelving banks, are also his favourite haunts; and he is pretty sure to be found under some old decaying wooden jetty, the foundation of which has been bared by the rapid rushing of the river. Some say his flesh is white and delicate; others, that it is coarse and insipid; so who shall decide? One thing is certain, in eating him, if not very careful, you will run the risk of choking yourselves with his bones. He will live four or five hours after he is taken out of the water. The barbel is in season from March to the end of October; and the likeliest time to make sure of him is late in the evening, or very early in the morning, especially after rainy weather; he may also be taken in the afternoon, but he does not then bite so freely.

Sometimes he will remain stationary for hours in one spot; and, although it is considered unfair, he is then often taken with eel-spears, landing-hooks, and similar instruments. But no matter how coarse he eats, he gives the angler good sport, not unmixed with fear for the safety of his tackle; for when of large size, he is an extremely strong and crafty fish, and will use every expedient to get off the hook, or else snap the line, which, unless the angler exerts his skill, he will certainly achieve. Before fishing for barbel, throw in plenty of ground-bait, and continue to do so at intervals; the best bait is one made of soaked greaves, bran, and clay, mixed together in balls about the size of an egg; a quantity of worms chopped into pieces make also a very good bait; salmon or trout spawn, maggots, and a paste composed of sheep's suet, cheese, and honey, mixed together, are likewise very tempting to this fish. The barbel being a very sharp and quick biter, you must strike smartly the moment you see a nibble, then let him run some distance before you turn him; keep him away from weeds, strive to get him into deep water, play him until he has lost all his strength, and then haul to land. In the Thames, barbel are usually fished for from punts or boats; a strong rod is necessary, with running tackle, gut line, quill float, and a No. 7 or 8 hook; the bait should always touch the bottom of the stream.

THE EEL.

The eel, as every boy knows, is a slippery subject, and unless properly managed, gives more trouble to get the hook out of him, and to keep him from twisting the line into knots, after he is caught, than to catch him; and, cruel as it may seem, the only way to save your tackle after he is trolled out is to set your foot on him and then cut off his head. There are three varieties of eel common to this country—the sharp-nosed, the broad-nosed, and the snig-eel, all differing in the shape of their noses. In length, the eel is found from one to three feet, and often longer; they vary in colour, some being a dark olive brown on the back, or light brown, but generally of a silvery whiteness on the belly. The head is flat; the eyes very near the mouth, iris reddish; the gill opening a good way back, close to the fin, and the lower jaw the longest. The eel leaves our rivers to spawn in the sea, and neither mill-dam nor floodgate can retard its progress at this season. The young first appear on our coast, and at the mouths of our rivers, in March and April, when they are about half an inch or an inch in length, and always swim in a double column, one close to the other, and thousands in each column. They love muddy and still waters, and are in season all the year round. There are several methods of taking them—

viz., by rod and line, night and dead lines, sniggling, bobbing, and trimmer fishing; and the most alluring baits are wasp-grubs, maggots, and small red worms. If you use a rod, the line must be either of strong gut or twisted hair, and the hook No. 8 size. Let the bait touch the bottom, and when you perceive a bite, allow the float to remain for a moment under water before you strike. When  using the dead line, which should be of whipcord, a bank runner must be employed; five or six hooks should be put on the line about nine inches asunder, and they should be baited with small fish or lobworms, as the latter remain on the hook alive for a considerable time.

For sniggling, the line must be either of platted silk or whipcord, and instead of a hook a stout worsted needle should be fastened by its middle to the line. A large marsh or small lob worm, tough and well scoured, is the best bait for this species of fishing, and when you bait your needle you must thrust its point into the worm's head, and draw it through the body of the worm until the latter completely enshrouds it. When you go out sniggling you should carry the line on a winder in your hand, and search for the fish near floodgates, wharves, bridges, piles, holes in the banks of rivers, ponds, and canals, and also in ditches, and amongst osiers and willows. You must put the bait into the lurking-holes by means of a stick with a forked head, and when you find that the bait is taken, by the line being pulled further into the hole, give the fish a few seconds to gorge, and then strike smartly, which will instantly cause the hook to fall across in his stomach; then hold the line fast and pull it towards you. Bobbing for eels is thus practised: a large quantity of marsh-worms should be procured, and as many as will make a bunch about the size of a turnip, strung on worsted by passing a needle through them from head to tail, and fastening them on your line, so that all the ends may hang level; affix in the middle of the bunch a leaden plummet of a conical form, and then tie the whole to a stout rod or pole. Having thus prepared your material, cast your bait softly into the water, and move it gently up and down until you perceive by the jerks on the line that the eels are attracted by the bait; then draw the line very steadily to the surface, and land it with all possible expedition. During warm weather the shallow parts of the stream are the most likely haunts of these fish, and where most sport may be obtained.

The eel-spear is also commonly used in taking eels, but as this requires more strength than art, it needs no description, for it is only used generally by fishermen who look for profit, and care nothing about the sport. But the great bulk of the eels caught in this country are taken in traps set in the weirs of rivers, when the eels run in the floods and freshes which are so common to all our rivers. Though he is so delicious when cooked, very few anglers really care about catching him, nor are they ever surprised at finding him on the hook, let them be fishing for whatever they may, with small bait.

For trimmer fishing, a double hook tied to gimp must be fastened to a line about fifteen or twenty yards in length, at about a foot above the hook, a wine-bottle cork should be firmly fixed on the line, and a bullet about two feet above the cork. When baiting your hook—suppose with a gudgeon—take your baiting-needle and hook it to the loop of the gimp on which the hook is fastened, thrust the point of the needle under the skin of the fish near the backbone, beginning about a quarter of an inch from the head, and pass it very carefully along between the skin and the flesh until within an inch of the tail, and then draw the needle and gimp out so that the hooks may come to the place where the needle was inserted; fasten off the loop of the gimp to the line, and the bait is then ready, as shown in the illustration. If the bait is put carefully on the hook, it will live some hours in the water. Select a place free from weeds, hold the line in your left hand, and a forked 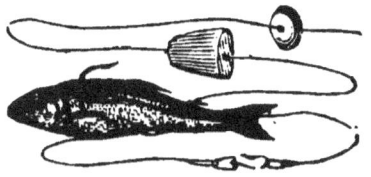 stick in your right; put the stick under the line just above the bullet, and by giving it a jerk you may throw the bait into any part of the river you please; then fasten the line either to a peg stuck firmly on the ground, or to a bank runner—which last is decidedly the best plan, as it prevents the string from entangling—a disaster that frequently occurs in the ordinary method. The night lines must be of strong small cord, about ten yards in length, and the baits small dead roach, gudgeon, or dace; some anglers, however, use worms and frogs. If you employ a fish, you must put it on a double hook, such as are sold at the shops, tied to gimp, wire, &c., and called eel hooks, in the manner described for trolling for pike; then put the hook on the line, and about two feet above the hook a bullet, which serves to steady both line and hook. The line should then be cast into the stream, and be securely fastened to a peg driven into the bank. Sometimes a number of these lines are fastened to a stout cord about two feet apart from each other, which forms what is termed a chain line; then a brick is affixed to one end of it, while the other is tied to the peg as before described, when the line is thrown all its length into the river and left there all night; other fish are often found on these night-lines.

As we have before remarked, salmon-fishing requires too much skill for boys, so does parr, common trout, lake trout, char, and grayling-fishing, and as these are neither common fishes, nor found in our common rivers, but some in Scotland, some in the Lake district, and others in out-of-the-way places, we must refer our young readers to Mr. Blakey's work on Angling, published at the low price of one shilling; or if all that can possibly be known is sought for, to that admirable work by Stonehenge, entitled, the "Manual of British Rural Sports," where everything appertaining to fishing, racing, hunting, shooting, coursing, boating, &c. &c., may be found.

SALT WATER ANGLING.

Many kinds of fish may be caught at the mouths of some of our rivers, when the tide is running in from the sea, such as plaice, whiting, small codfish, turbot, haddock, and others, which will readily seize a bait, and may be angled for from piers and projecting rocks, and even mackerel may be taken from similar places, during the time they are in season. For this kind of angling, a good strong rod, stout, well-leaded line, large cork float, and good-sized hook, are requisite. When fishing at the mouths of rivers with gentles, well-scoured red worms, or shrimps, as baits, you may also take eels, flat-fish, and smelts; when in a boat, a short distance from land, two or three large red worms, a small raw crab, or a mussel, or a little bit of whiting, will prove very serviceable baits. A piece of scarlet cloth will tempt mackerel, and to ensure success, it is necessary to let your bait hang about eighteen inches below the surface of the water, or even lower, if you can allow it.

NATURAL FLY-FISHING.

Natural fly-fishing, usually termed dibbing or dapping, consists in fishing with the living flies, grasshoppers, &c., which are found on the banks of the rivers or lakes where you are fishing; it is practised with a long rod, running tackle, and fine line. When learning this system of angling, begin by fishing close under the banks, gradually increasing your distance until you can throw your live-bait across the stream, screening yourself behind a tree, a bush, or a cluster of weeds, otherwise you will not have the satisfaction of lifting a single fish out of the water. In rivers where immense quantities of weeds grow in the summer, so as almost to check the current, you must fish where the stream runs most rapidly, taking care that in throwing your line into those parts you do not entangle it amongst the weeds. Draw out only as much line as will let the fly just touch the surface, and if the wind is at your back, it will be of material service to you in carrying the fly lightly over the water. In such places the water is generally still, and your bait must if possible be dropped with no more noise than the living fly would make if it fell into the water. Keep the top of your rod a little elevated, and frequently raise and depress it and move it to and fro very gently in order that the fly by its shifting about may deceive the fish and tempt them to make a bite. The instant your bait is taken, strike smartly, and if the fish is not so large as to overstrain and snap your tackle, haul it out immediately, as you may scare away many while trying to secure one. There are very many baits which may be used with success in natural fly-fishing, of which, however, we shall content ourselves with enumerating some of the most usual and useful only.

The red copper-coloured beetle is an extremely good bait if the outer hard wings are clipped, and the insect hung with its feet downwards.

Wasps, hornets, and humble bees, are esteemed good baits for dace, eels, roach, bream, chub, and flounders; they should be dried in an oven or over the fire, and if not overdone, they will keep a long while.

In March, the blue dun and cow-dung flies make their appearance, and may be used throughout the year. The March brown fly appears about the same time, but is out of season at the end of April; it is a capital bait, and it kills most from eleven till three.

In April, the green tail and gravel flies come out; they are soon out of season, the former continuing not more than a week, and the latter about a fortnight. The black gnat, which continues till the end of May, and the stone fly, complete the list for April.

In May, the increasing warmth of the weather brings more insects forward; accordingly, the green drake, the grey drake, the fern, hazel, ash, orl, little iron blue, and yellow sally flies, form the bill for the month. The two first flies appear much about the same time, and are most excellent baits in trout fishing; they continue in season about a month; and are very plentiful on sandy, gravelly streams. The fern and ash flies continue till September; the hazel, yellow sally, and little iron blue flies, for a month, and the orl fly for about two months.

In June, the white gnat, cock-tail, gold-spinner, governor, blue gnat, whirling dun, hare's ear, and kingdom flies. The gold-spinner, governor, and kingdom flies continue till August; the blue gnat for about a fortnight, and the other flies during the summer.

In July, August, and September; in the first named month, the red ant; in the second, the whirling blue; and in the last, the willow fly; they continue in season till fishing is over.

Ant flies may be procured from June till September in their hills; they are never-failing baits for chub, roach, and dace, if you let your hook hang about six inches from the bottom of the stream.

The great white moth, which can be obtained in the summer evenings in gardens, on trees and shrubs, is a serviceable bait when dibbing for roach in the twilight.

The hawthorn fly makes its appearance on hawthorn trees, when the leaves are beginning to sprout; it is a dark coloured fly, and is used as a bait for trout, though there are many other fish that will snap at it eagerly.

The bonnet fly, which frequents standing grass, is an extremely good bait for chub and dace.

Common flies are, by some anglers, reckoned the best baits for dace and bleak; two or three of them at a time should be put on a No. 10 hook for dace, and one on a No. 12 hook for bleak.

Ant flies must be kept in bottles, in some of the earth from which they are taken. Common flies may be kept in a bottle, but the most convenient natural fly-holder is a horn bottle made in a conical form, having a wooden bottom pierced with a multitude of small holes to admit air, and which apertures must be so small that the minutest fly you employ cannot escape through. The apex of the cone should be stopped up with a cork, so that by uncorking it you may take out your baits easily without losing any of them.

ARTIFICIAL FLY-FISHING.

Artificial fly-fishing consists in the use of imitations of these flies and of other fancy flies, and is unquestionably the most scientific mode of angling, requiring great tact and practice to make the flies with neatness and to use them with success, and calling forth as it does so much more skill than the ordinary method of bottom fishing, it merits its superior reputation. It possesses many advantages over bottom fishing, but at the same time it has its disadvantages; it is much more cleanly in its preparations, inasmuch as it does not require the angler to grub for clay and work up a quantity of ground-baits, and is not so toilsome in its practice, for the only encumbrances which the fly-fisher has are simply a light rod, a book of flies, and whatever fish he may chance to catch; but there are several kinds of fish which will not rise at a fly, and even those that do will not be lured from their quiet retreat during very wet or cold weather. It would be as well if the young angler could go out for some little time with an old experienced hand, to observe and imitate his movements as closely as possible; but as many of our readers will not, in all probability, be able to enjoy such an advantage, we subjoin some instructions by which they may pursue this interesting branch of angling.

INSTRUCTIONS FOR ARTIFICIAL FLY-MAKING.

The artificial flies sold at the fishing-tackle shops are manufactured so skilfully and naturally, that in our opinion the young angler would find it much more to his advantage to purchase them ready-made than trouble himself with their fabrication; but for the guidance of those who would rather fish with a fly of their own contriving, we proceed to describe the proper method of making some few out of the many in use as plainly and as concisely as we can, candidly stating, however, that when they have done their very best—until after long practice—they will manufacture a very inferior article to that sold at the first-rate fishing-tackle shops.

The mere enumeration of the various articles necessary for artificial fly-making will appal a timid boy, while a patient, persevering boy will delight in the difficulties. They are feathers of no end of birds, and fur, and hair of a great variety of quadrupeds, also sewing silk of different tints and thicknesses, and gold and silver twist. Amongst the tools, a pair of fine-pointed scissors and a pair of small pliers must be enumerated; wing-picker or pointer, pair of fine spring-forceps, silks of all kinds and colours, wax, spirit-varnish and brush; and these should be kept in one of the cases which are made expressly for the purpose. Before you begin your task, see that you have all the materials you imagine will be required in readiness and close at hand, and also try the strength of the gut; then take the hook in the left hand, sticking the hooked end in a cork which is made fast; wind some silk round the bare hook two or three times, and lay the fine end of the gut on the under side of it, and beginning at the bend, wind the silk three or four times round both the gut and the hook, fasten in the hackle, and continue winding on the silk until you reach the end of the hook, when you must form the head of the fly by turning the silk back and winding it several times round. Next twist the dubbing on the silk, and wind it on the hook for nearly half the intended length of your fly, and fasten it off; when you have wound enough of the feather upon the hook, you should hold the remainder under your left thumb, and with a needle pick out the twisted and entangled fibres; continue twisting the silk and dubbing over the end of the hackle until you make the body of the fly the proper length, and then fasten off.

In making a winged fly, fasten the hook on the gut in the manner above described, take the feather intended for the wings and place it on the upper side of the shank with the roots turned towards the bend of the hook, and fasten the feather securely down by twisting the silk over it; clip the root ends close with your scissors, and with a needle divide the wings as evenly as possible, passing the silk two or three times between them, so as to make them take their proper position. Carry your silk towards the bend of the hook to about the length which you intend your fly to be, and fasten it there, lay on your dubbing, and then continue winding the silk up towards the wings; put the hackle in for the legs, and wind it so nicely under the wings, that the ends of the cut fibres may be quite hidden, and then fasten the silk off above the wings. When gold or silver twist is used, it should be fastened to the lower end of the body before the dubbing is put on. The fly at the end of the line is usually termed "the stretcher," and the others "droppers." The first dropper should be put on the line about a yard above the stretcher, and the second about three quarters of a yard from the first; they should be made on separate pieces of gut about four inches in length, for the purpose of being taken off at will.

ARTIFICIAL FLIES.

There being upwards of a hundred different kinds of flies suitable to this species of angling, a full description of the method of making each would far exceed our limits; we shall, therefore, only describe a few of the commonest, for a complete work on artificial fly-making would of itself make a large volume, and such a work may be had written by Mr. Blakey, who is said to be one of the best fly-makers in Great Britain. The cow-dung fly may be used from the first of April, and will kill till September. Its wings should be made of a feather of the land-rail, its body of yellow camlet mingled  with a little fur from the brown bear, and its legs of a ginger hackle; its wings should be trimmed flat. The blue dun is an excellent fly during March and April, and should be used in the middle of the day. Make its wings of a starling's feather, body of blue fur from a water-rat, mixed with a little yellow-coloured mohair, and its tail, which is forked, of two fibres from the feather which you use for the wings. The black gnat makes its appearance about the latter end of April, and will be found useful till the close of May. A black ostrich's harl must be used in making the body of this fly, a starling's feather for the wings, and very fine black cock's hackle for the legs; it should be trimmed short and thick. This fly is reckoned a good killer when the water is rather low. The violet fly is also used in April; it is made of light dun-coloured bear's hair, mixed with violet stuff, and winged with the grey feather of a mallard. The stone fly, which may be used with success during May, especially in the mornings, has a body of red mohair ribbed with gold or yellow silk; tail, two long fibres from a red cock's hackle; red cock's hackle also for the legs, and a mottled feather from a hen pheasant, or grey goose wing feather hackle for the wings, which must lie flat.

The green drake, or May fly, is perhaps the best fly that can be procured for trout fishing. Its wings should be made of the light feather of a grey drake, dyed lemon colour, its body of yellow-coloured mohair, neatly ribbed with green silk, head of a peacock's harl, and its tail of three long hairs from a sable muff. The yellow sally is an approved fly from the early part of May to the end of June; its body must be made of yellow unravelled worsted, mixed with some fur from a hare's ear, and its wings of a hackle dyed yellow; the wings of this fly must lie flat. The grey drake appears about the same time as the green drake, which in shape it very closely resembles, and is a serviceable fly from three o'clock till dusk. Its body must be made of pale dun-coloured mohair, the tail of two fibres from the mallard's back, legs of a brown or ginger cock's hackle, and wings from the grey feather of the mallard's back, undyed. The oak fly, down-looker, or ash-fly, is usually found on oak and ash trees, during May and June, with its head pointing downwards towards the roots of the trees. Its wings, which must lie flat, should be made with a wing feather from a woodcock, its body of dun-coloured fur mingled with brown mohair, and its head of fur from a hare's ear. The purple fly made of purple wool mixed with light brown bear's hair, and dubbed with purple silk, is useful during June and July.

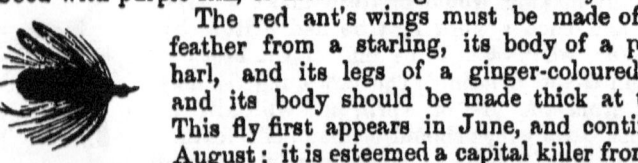

The red ant's wings must be made of a light feather from a starling, its body of a peacock's harl, and its legs of a ginger-coloured hackle, and its body should be made thick at the end. This fly first appears in June, and continues till August; it is esteemed a capital killer from eleven in the morning till six in the evening.

The dark alder fly, in May and June, is a great favourite; it may be imitated by a dark shaded pheasant's wing, black for its legs, and a peacock's harl ribbed with silk hackle for the body. The fern fly also appears in the early part of summer; its body is dubbed with the wool from a hare's neck and its wings made of a darkish grey mallard's feather

The palmer is a most killing bait during the summer, and its body should be made to resemble a hairy caterpillar, with black spaniel's fur on the harl of an ostrich feather, wrapped over with a red hackle from a game cock. The wasp fly, made of brown dubbing or the hair of a black cat's tail ribbed with yellow silk, is an excellent fly during the month of July; and about the same  time the orange fly is in vogue, its wings are made of a feather from a blackbird's wing, and its body of orange-coloured crewel or wool. The whirling blue makes its appearance in August, and is a good bait till the end of the season; its wings must be made of a feather from a common tern, its body of light blue fur mingled with a little yellow fur, and its legs of a light blue hackle.

The whirling dun is also a summer fly, and its wings are made of a snipe's feather, its body of blue fur wrapped with yellow silk, its legs of a blue cock's hackle, and its tail of two hairs from a light coloured muff. The late badger-fly is also serviceable in August; it is made of black  badger's hair whipped with red silk, and winged with a darkish grey mallard's feather. Imitations of the house and blue-bottle flies are taken greedily in August; they are also particularly killing after floods in autumn. The willow fly appears in September, and is the best bait during that month and the remainder of the season; its body must be made of blue squirrel's fur, with a little yellow mohair intermixed, and its wings of a dark grizzled cock's hackle.

Our limits compel us to close this brief sketch of artificial fly-making, for were we to describe the way to manufacture every fly at which a fish will rise, our labours would be almost interminable; it would be also an unnecessary task, as the same methods, varied only as to colour and material, answer for every kind of fly; and if the young angler can fabricate those we have endeavoured to describe, he can certainly imitate any other he may chance to require.

CASTING THE LINE.

The single rod for artificial fly-fishing should be from eleven to thirteen feet long, light and flexible, and one which you can wield with ease. Raise your arm, and swing the rod back, without effort, so as almost to describe a circle round your head, and when the line has reached its full extent behind you, throw it forwards, taking great care in the movement, else when you have a fly on the line you may, perchance, jerk it off, particularly if you attempt to make the forward move ere the line has reached its full circuit. In order to acquire a good style of throwing, and a correct eye for measuring distances, it it proper to practise at first at a short length only, without a fly on the line, and when you can throw to a moderate range with a tolerable degree of certainty, one fly may be put on, and practised with awhile; and as still further expertness is gained, two or three may be employed; it is also a good plan to fish in rapid streams until a dexterous mode of casting the fly is arrived at. When casting the line use your utmost endeavours to drop the fly lightly on the surface of the water; for the more skilfully you throw the fly, the greater your chance of success, and this desirable adroitness can only be acquired by constant practice. When you perceive a rise, throw your fly a little above the spot, and let it drop gradually down the stream, and directly the bait is taken, for which you must keep a sharp look out, strike quickly, or else the fish will discover its artificial character and refuse it. When you have hooked a fish, run him down the stream, play him very cautiously, keep his head up, and at the same time draw him by gentle force towards you. Keep your back to the wind if possible, as you can then stand further out of the fish's sight, and so angle on both sides of the river, if it is not a very broad one; and if the sun is shining, stand with your face to it, that your shadow may not be cast upon the water; if the day is so calm that the very reeds are motionless, then keep as far away from the brink as you can, as fish are extremely quick-sighted, taking an alarm and vanishing in an instant at the slightest appearance of danger. Generally speaking, the best time for fly-fishing is when the day is overcast and gloomy after a beautiful clear night, or when a light breeze just agitates the surface of the stream; and if the wind is from the south or west, and the water turbid from recent heavy rains, it is all the better for the angler. Fish every yard of water likely to afford sport, and keep your fly continually in motion, that it may appear to be a natural one. The list of natural flies, with the months in which they usually appear, appended to the article on natural fly-fishing, will serve as a guide to the proper times for using artificial ones, as fish seldom rise at imitations of flies not naturally in season.

A downright good fly-fisher will throw his fly as gently on the water as if it had alighted there of its own accord. A bad fly-fisher will have two or three feet of line touching the water; a good one not as many inches.

MONTHLY GUIDE FOR BOTTOM FISHING.

JANUARY.—Chub, pike, and roach are the only fish that can be taken in this month; the middle of the day is the most seasonable time, provided the water is tolerably clear, and free from ice.

FEBRUARY.—Towards the latter end of this month, when the weather becomes somewhat milder, carp, gudgeons, and minnows may be taken, as well as pike, chub, and roach. The middle of the day is the most favourable time, and fish in eddies near banks. The perch spawns either in this or the next month.

MARCH.—In this month, minnows, roach, chub, gudgeons, tench, carp, and trout form the bill of fare. Smelts, bleak, pike, and perch spawn. In this month, also, the middle of the day is the best for angling.

APRIL.—The increasing warmth of the weather brings also increase of sport to the patient angler, and tench, perch, trout, roach, carp, gudgeons, flounders, bleak, minnows, and eels reward his toil.

Barbel, pike, chub, ruffe, and dace are out of season, his being their spawning time.

MAY.—In this month, perch, ruffe, bream, gudgeons, flounders, dace, minnows, eels, and trout may be taken. Carp, barbel, tench, chub, roach, and bleak spawn.

JUNE.—Roach, dace, minnows, bleak, gudgeons, eels, barbel, ruffe, perch, pike, and trout are in season. Carp, tench, bream, and gudgeon spawn about this time.

JULY.—The list is still tolerably comprehensive: trout, dace, flounders, eels, bleak, minnows, pike, barbel, gudgeons, and roach affording good sport. Bream and carp spawn.

AUGUST. — In this month, barbel, bream, gudgeons, roach, flounders, chub, dace, eels, bleak, minnows, pike, ruffe, and perch bite freely.

SEPTEMBER.—Roach, gudgeons, dace, chub, eels, tench, bleak, minnows, barbel, bream, ruffe, pike, trout, and perch are in season.

OCTOBER.—Tench, gudgeons, roach, chub, dace, minnows, bleak, pike, trout, and grayling are the principal fish in season this month; trolling or bottom fishing for chub and roach may be practised with success; fly-fishing is generally over about this time.

NOVEMBER.—This month's list is limited; roach, pike, chub, trout, and grayling being the only fish now in season. Roach and chub get into deep waters and remain there till spring.

DECEMBER.—When the weather is propitious, pike, roach, and chub may sometimes be taken, but nearly all other fish have retired to their winter retreats, to screen themselves till the return of spring.

The proper time of spawning, as well as that when fishes are in season, are given more fully under the heading of every fish as we have described them separately. Fishes do not, however, always spawn in the same months, as a very cold spring will throw the March spawners into April, and the latter very often into May.

FISHING STATIONS NEAR LONDON.

The river Thames contains a vast number and a great variety of fish, and the angling stations upon it are very numerous.

BATTERSEA BRIDGE was, a few years ago, a favourite spot of the anglers, but the steamboats have nearly spoiled all fishing until Putney Bridge is reached.

At BRENTFORD there is a place called the Aits.

RICHMOND is a spot much frequented, but the water is preserved 683 yards from the pier eastward.

TWICKENHAM long deep preserved water extends 410 yards.

TEDDINGTON LOCK, however, is only a mile and a half further up, and here there are plenty of fish.

KINGSTON is also a good spot for rod-fishing.

THAMES DITTON is a good spot, but much of it is preserved.

HAMPTON COURT and HAMPTON bear a good reputation, but here the preserved waters again intervene.

WALTON, on the Surrey side, has some splendid water here, only a portion of which is preserved.

SHEPPERTON has some famous fishing on both sides the river.

WEYBRIDGE, CHERTSEY BRIDGE, LALEHAM, and STAINES are within an hour's ride by rail, and there is good fishing to be had at all these stations, which ought to be visited when the London boys have holidays.

We will now take a brief survey of the

TRIBUTARY STREAMS OF THE THAMES,

And a few other places near to and around London.

The LEA is a favourite river, commencing at TEMPLE MILLS, which is free water, while at the WHITE HOUSE, Hackney Marsh, the liberty of fishing must be paid for, either by the day or year; but a portion of the river above Lea Bridge is free.

TOTTENHAM MILLS has a subscription water.

BLEAK HALL is a spot much frequented; the terms are two guineas a year.

WALTHAM ABBEY is a famous place for fish, but permission must be obtained.

BROXBOURNE, PAGE'S WATER, and the RYE HOUSE are all good places for bottom-fishing.

For the NEW RIVER tickets must be obtained.

There is capital bottom-fishing in all the LONDON DOCKS, and there is no difficulty in obtaining a ticket.

SURREY CANAL contains roach, perch, and eels.

PADDINGTON CANAL, chub, eels, gudgeons, carp, pike, and roach.

On HAMPSTEAD HEATH and CLAPHAM COMMON there are fish-ponds.

HORNSEY WOOD HOUSE has a lake in the gardens.

The RODING, which runs into the Thames at Barking, produces an abundance of fish. There are many favourable spots for angling in this river, at Abridge, Woodford Bridge, Loughton, Ilford, Wanstead, and Barking.

In the MOLE the angler will find the best sport near Esher, Leatherhead, Cobham, Dorking, or Reigate.

In the WANDLE, at Mitcham, Merton, Carshalton, Wandsworth.

The RAVENSBOURNE, in Kent, contains good fish.

On CHISELHURST COMMON, in Kent, the large pond near the Queen's Head Inn is the best spot.

A mile to the east of Shooter's Hill, there are some ponds on a common near the road-side, free to all anglers.

The river WEY, in Surrey, contains good fish.

DAGENHAM BREACH, in Essex, is preserved.

At STANMORE, in Middlesex, ten miles from London, there are two or three ponds on the common, and between these ponds and Stanmore Priory, about a mile distant, is a very fine piece of water called the LONG POND.

SNARESBROOK, in Essex, is a subscription water.

In the COLNE, near Uxbridge and Denham, fine trout abound; but you must obtain leave to fish.

A LIST OF SOME OF THE MOST CELEBRATED RIVERS OF ENGLAND, WITH THE FISH WHICH MAY BE FOUND IN THEM.

Having now described the principal rivers, &c., in the neighbourhood of London, we shall proceed to notice briefly a few others in various parts of the kingdom.

The "stately SEVERN," which rises in Montgomeryshire, and after running through part of Shropshire and Worcestershire, passes Gloucester, and discharges itself into the Bristol Channel, near King's road, is a most excellent salmon and trout stream, and likewise abounds with other fish.

The TRENT first appears in Staffordshire, runs the whole length of Nottinghamshire, and falls into the sea south-east of Hull. It is well stored with pike, eels, carp, bream, barbel, chub, perch, grayling, roach, and flounders. Several minor streams run into it, such as the DOVE, the SOUR, the IDLE, the LEANE, &c., all of which are well stocked with trout and grayling.

The STOUR, which rises in Kent, and empties itself into the sea near Sandwich, abounds with trout, eels, roach, &c.

The MEDWAY also takes its rise in, and passes through Kent; it empties itself into the sea at Sheerness, and is well stocked with eels, perch, pike, flounders, and a few other fish. Salmon may also be taken in this river, but they are far from plentiful.

The ITCHIN, which rises in Hampshire, contains trout, large eels, and many other fish. It runs into the sea at Southampton.

The ISIS and the CHARWELL, near Oxford, afford perch, roach, and pike in great plenty. In the Charwell, a very scarce fish called the reid, or feuscale, may be found.

The KENNET, which rises near Marlborough, in Wiltshire, and falls into the Thames near Reading, is an excellent trout stream. In the vicinity of Newbury, the right of fishing in this river is vested in the householders of that town.

The river TEST, in Hampshire, is one of the finest trout streams in England; grayling may also be found in it, in the neighbourhood of Houghton. This river runs into the Southampton Waters at Redbridge, a short distance from Southampton.

The EX rises in Somersetshire, passes Tiverton and Exeter, and discharges itself into the sea at Exmouth. It is well stocked with salmon, trout, eels, &c., as are also many small streams which run into it.

The WYE, which rises in Montgomeryshire, passes Hereford and Monmouth, and falls into the Severn below Chepstow, is stored with trout and grayling.

The OUSE abounds with pike, bream, eels, perch, &c.; this river rises in Oxfordshire, and after passing by Buckingham, Bedford, Huntingdon, and Ely, empties itself into the sea at Lynn, in Norfolk.

The CAM rises in Cambridgeshire, passes by Cambridge, and finally blends itself with the Ouse; it is well stored with carp, pike, roach, eels, perch, &c.

There are several MERES, or pieces of water in the vicinity of this river, stocked with tench and other fish.

RAMSAY MERE, in Huntingdonshire, is famous for its eels and pike.

In the TAMAR, which divides the counties of Cornwall and Devon, more salmon may be found than in any other river in the West of England.

The LOWER AVON, near Salisbury, contains grayling.

The little TEME and the CLUN, near Downton, in Shropshire, abound with trout and grayling. The Teme contains trout principally; the Clun, both trout and grayling.

LAWS RELATIVE TO ANGLING.

By an Act of Parliament passed in the 7 & 8 George IV. for consolidating and amending the Laws relative to Larceny, &c., it is provided, "That if any person shall unlawfully and wilfully take or destroy any fish in any water which shall run through or be in any land adjoining or belonging to the dwelling-house of any person being the owner of such water, or having a right of fishery therein, every such offender shall be guilty of a misdemeanour, and, being convicted thereof, shall be punished accordingly; and if any person shall unlawfully and wilfully take or destroy, or attempt to take or destroy, any fish in any water not being such as aforesaid, but which shall be private property, or in which there shall be any private right of fishery, every such offender being convicted thereof before a Justice of the Peace, shall forfeit and pay, over and above the value of the fish taken or destroyed (if any), such sum of money not exceeding five pounds, as to the Justice shall seem meet; provided always that nothing herein before contained shall extend to any person angling in the day-time; but if any person shall by angling in the day-time, unlawfully and wilfully take or destroy, or attempt to take or destroy, any fish in any such water as first mentioned, he shall on conviction before a Justice of the Peace, forfeit and pay any sum not exceeding five pounds; and if in any such water as last mentioned, he shall, on the like conviction, forfeit and pay any sum not exceeding two pounds, as to the Justice shall seem meet; and if the boundary of any parish, township, or vill, shall happen to be in, or by the side of any such water as is herein before mentioned, it shall be sufficient to prove that the offence was committed, either in the parish, township, or vill named in the indictment or information, or in any parish, township, or vill adjoining thereto.

"And be it enacted, that if any person shall at any time be found fishing against the provisions of this Act, it shall be lawful for the owner of the ground, water, or fishery, where such offender shall be so found, his servants, or any person authorized by him, to demand from such offender any rods, lines, hooks, nets, or other implements for taking or destroying fish, which shall then be in his possession; and in case such offender shall not immediately deliver up the same, to seize and take them from him for the use of such owner; provided always, that any person angling in the day-time against the provisions of the Act, from whom any implements used by anglers shall be taken, or by whom the same shall be delivered up as aforesaid, shall by the taking or delivery thereof be exempted from the payment of any damages or penalty for such angling."

By another Act passed in the 7 & 8 George IV., it is provided, that "if any person shall maliciously in any way destroy the dam of a fishpond or other water, being private property, with intent to take or destroy any of the fish in the same ; or shall maliciously put any noxious material in any such pond or water with intent to destroy the fish therein, such offender shall be guilty of a misdemeanour, and be punished accordingly."

The provisions of these Acts do not extend to Scotland and Ireland.

HINTS FOR ANGLERS.

It is generally understood that when two or three persons are angling in the same stream, there shall be a distance of thirty yards between them.

If the learner wishes to become a *complete* angler, he must use fine tackle, as the skill and care which such tackle requires will soon make him a master of the art.

When the tackle breaks, the angler must not mourn over the accident, but do his best to remedy it, by speedily repairing the damage, and resuming his sport.

The angler must wear strong boots or shoes, and keep his feet dry, unless he wishes to become an interesting invalid.

And if he values his health, he will abstain from drinking water out of rivers or ponds when he is in a perspiration, or feels parched with thirst.

If the weather is very cold, or the wind sets very strongly from the north or east, the angler will meet with but little sport. Heavy showers of rain or hail, and thunderstorms, are likewise extremely prejudicial to his amusement ; and as in the winter months few opportunities are afforded for the exercise of his talents out of doors, he should, while snugly screened from the pinching blasts, attend to all the little repairs which may be necessary to his various appurtenances.

BRITISH SONG BIRDS.

THE sweetest music of Nature is the singing of those little angels of the trees—the birds. It gladdens the heart to hear their "wood notes wild" ringing through those great cathedrals the woods, whose tall green pillars are reared by the wonder-working hand of God. Their song seems to make the sunshine brighter; and we have often fancied we could see the golden rays darting, and flashing, and keeping time to their warblings. No doubt when long-haired Eve wandered through Eden, and Adam wove for her a bower of blushing roses, that the birds sang the self-same tunes which they do now, and that the angels often listened to them when they "walked in the garden in the cool of the day."* In all ages, in all countries, as far back as we have any records of time, the singing of birds has given delight to mankind; and they must many a time have gladdened the heart of Noah and his family when he sailed over the wide waste of waters in the ark; and the children that were with him would listen and recal the green nestling places, then deep down beneath the waves, where the sweet-voiced birds built and sang.

Everything about a singing bird is beautiful; the very nest it builds, the eggs it lays, are all objects of beauty. As for their songs, they seem sent to gladden the hearts of mankind, as the flowers delight the eyes of the beholders. Even the cheerful trills of the little captives when caged in their wired prisons in some dingy apartment

---

* "They heard the voice of the Lord God walking in the garden in the cool of the day."—Genesis iii. 8.

in a city, are often the source of unmingled pleasure to the plodding money-getting citizen; and the poor weaver, though plunged in the utmost depths of penury, feels pleasure, whilst plying the busy loom, in teaching bulfinches to warble national airs, and in training goldfinches to perform their interesting little tricks. The humble peasant also finds amusement in teaching his starling to talk while hanging in its wicker cage upon "the woodbine arbour," and often a country cobbler may be seen sitting near the window of his snug roadside cottage, progressing cheerily with his work, and ever and anon pausing to listen with rapture and pride of heart to his blackbird piping some heart-cheering ditty or plaintive love song which he has taught it.

BIRD CATCHING.

Boys residing in London, or any large town, may always procure good, healthy birds, of strong musical powers, by applying to respectable bird-fanciers; whilst those who live in the country must either take the young from their nests and rear them, or use various contrivances in which to ensnare them; and there are, perhaps, few things which afford such an inexhaustible fund of amusement to country lads as bird-catching. In the budding days of spring, the sunny hours of summer, the sombre autumn, and the chill, piercing days of winter, they may set their cunningly-contrived nets and exert their skill in the construction of traps, the making of which will find them employment on winter nights by the fireside.

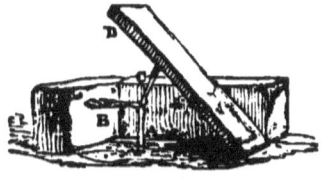

The common BRICK TRAP employed by the veriest children in bird trapping, is made of four bricks and a tile; two of the bricks are placed lengthways, parallel with each other, and the others are put at the ends; the tile acting as a cover, and to support it, a stump is driven into the ground, as represented at A, in the annexed engraving: upon this stump, one end of a forked twig B, is rested, and the other end is jutted close to the cross brick; on this forked twig, a short straight bit of stick C, is placed, and fragile as the support seems, the whole weight of the tile D, bears upon it. At the bottom and around the trap some seed should be scattered, and the apparatus is then complete. The instant a bird, attracted by the seed, alights on the forked twig, it jerks up, and of course displaces the slender prop of the tile, which instantly falls and encloses the little adventurer. This trap may be made of four bricks only, one brick being used instead of the tile, and so placed, that when it falls it will rest on the edge of the brick marked B, so as not to crush the bird.

The SIEVE-TRAP only requires a sieve, a piece of stick and a string, as shown in the engraving. When the ground is covered with snow, a space about the size of the sieve should be cleared, and some ashes sprinkled on the spot, then a few crumbs of bread, or red berries scattered on the ashes. The sieve should be propped up, over the clear space, by a bit of stick; and to the middle of the prop, a piece of fine twine, of sufficient length to reach the window or hiding-place, must be fastened; at the place of concealment the contriver takes his station to watch all the comers, and the instant he sees any birds settle beneath it, to enjoy the banquet spread before them, he jerks the string, the sieve falls, and those which are unfortunate enough to be under, are immediately trapped. You must then take a cloth or apron and draw it cautiously under the sieve, taking care not to elevate the sieve so as to allow the birds to escape, raise the ends of the cloth to the centre, and carry your prizes into the house.

HORSE-HAIR NOOSES, employed in the winter for catching larks, are thus made and set:—when the ground is covered with snow, take about a hundred yards of packthread, and at every six inches fasten a noose composed of two horsehairs twisted together, with a loop which will draw tight when the bird drags at it. When you set them, thrust little pegs into the earth every twenty yards, and fasten the packthread to them, so as to keep the nooses at about the height of a lark when running. Scatter a quantity of white oats on the snow, from one end of the line to the other; and when the birds haste to partake of the food, they will speedily get entangled in the nooses, from which, of course, they must be immediately taken out.

The SPRINGLE, which is rather a complicated affair, is one of the most efficient traps in use, and is constructed in the following manner:—Get a hazle switch of four feet in length, and to the taper end of it tie a piece of string of about fifteen inches in length, and near the end of this string fasten a catch or little piece of wood of half an inch in length, a quarter of an inch in breadth, and the eighth of an inch in thickness; and this piece of wood must be slightly bevelled off at one end, so as to adapt it to a notch in the "spreader." A very loose slip-knot must next be fastened to the end of the spring, below the catch, and then the spring part of the machine is complete, as shown at Fig. 1 in the illustration. The "spreader" is the next thing to be made, and for it a little switch of about eighteen inches in length is requisite; the small end of it must be bent back, and fastened to within an inch or thereabouts of the thicker end, so as to form a loop as in Fig. 2, and at the latter end a notch must be cut for the purpose of receiving the catch before mentioned. A "stump," Fig. 3, and a "bender," or pliant bit of switch, Fig. 4, each about eighteen inches in length, complete the springle.

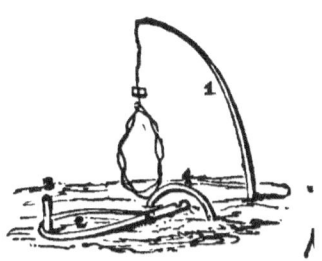

The method of setting it is the following:—Drive the stump, Fig. 3, into the ground, put the loop of the spreader over it, as Fig. 2, and at about the length of the spreader from the stump, thrust the ends of the bender firmly into the ground, as Fig. 4; then put the thick end of the springer into the ground at a little distance from the bender, as in Fig. 1, and bend it down until you can place one end of the catch upwards, on the outside of the bender, and then raise the spreader about an inch from the ground, and put the smaller end of the catch in the notch, by which arrangement the spreader will be held in its proper position, and the springer prevented from jerking up without some cause. Arrange the horse-hair slip-knot loosely round the spreader and stump, and scatter some seed inside it, and also sparingly outside, and for a little space around, to attract the birds to the more plentiful supply within the spreader; and the springle is then completely prepared, as shown in the illustration.

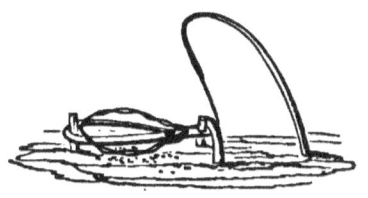

Its action is very simple, being as follows:—When a bird, attracted by the seed, perches upon the spreader, it falls with his weight, the catch is instantly freed, and in consequence, the springer flies up, ensnaring the poor bird in the slip-knot, either by the legs, neck, wings, or body. If the trappist wishes to take the birds alive, he must keep watch and ward within sight of the trap, so that as soon as one is imprisoned, he may run and take it out, else the poor bird will either strangle or beat itself to pieces in its vain endeavours to escape.

The CLAP NET is used for taking larks and other small birds, and the method is styled daring or doring. As the construction of this species of net is much too complex for our readers to attempt manufacturing any for themselves, and the sport being somewhat difficult, and only followed by men who make a living by bird-catching, we shall pass it over.

The NIGHTINGALE TRAP is of an oblong shape, about four inches in depth, with a perch or stick to support the top, which is so placed as to fall and secure the bird the instant he hops in to get at the bait.

The TRAMMELS NET is most generally made about thirty-six yards in length and six in breadth, with six ribs of packthread, the ends of which are fastened upon two poles each of sixteen feet long. The mode of using this net is for two persons to take it out on a dark night and drag it on the ground, touching the ground with it at intervals of every five or six steps, otherwise many birds would be passed over. The instant any fly up against the net, it is dropped, thereby securing all that are underneath. Many other birds which nestle on the ground, besides larks, are, as may be supposed, taken in this species of net.

BIRD-LIME is often recommended as a means by which birds may be taken, but it is a very ineffectual one. There are two ways of using it; the first is by smearing some small twigs with it, and laying them on the ground and scattering some crumbs of bread around them; the moment the birds observe the treat laid out for them, and alight, they get entangled with the twigs which adhere to their feet, and form a great inconvenience to them in flying away, even if they do not check their flight altogether. The other method of employing the lime is by smearing some over a hog's bristle, to the end of which a piece of bread has been secured; this is thrown upon the ground; a bird, little dreaming mischief, flies away with the piece of bread, and the bristle of course soon gets entwined around its wings, and brings it to the ground. This method is, perhaps, less effectual than the former, as the bird may fly some distance before it falls.

In taking young birds from the nest, great care is necessary, for if carried away when only stubbed or half-fledged, it is impossible to rear them by hand, as they require such constant feeding and attendance. The proper time for removing them is when the tail feathers begin to grow, for should they be taken at an earlier period, their stomachs will not support the change of food, and if at a later, in most cases it is difficult to make them open their beaks to take in food so novel to them. Some species of birds, however, are naturally so docile, that they may be taken at any age, and reared without difficulty.

As in our description of birds we have pointed out the best kind of cage adapted for each, it will only be necessary to give here a brief description of

THE BREEDING CAGE.

Breeding-cages are sometimes made single, sometimes double, and the size must of course be regulated by the convenience at the disposal of the young fancier. The top, front, and sides are usually wired, and the back o wood, but if the cage is double, both ends are then made of wood. Different drawers for food, and one at the bottom for readiness in cleaning, and glasses for water, should be furnished; the perches should be placed at various heights, and in the most convenient places. The door may be put according to fancy; in some cages it is at the side, as in the above representation. A small shelf should be fastened to the boarded back, and from the edge of the shelf a partition should be carried up to the top of the cage; this little shelf, with its partition, serves as a private chamber in which the birds may construct their nests, and two small boxes, or rather trays, are put inside for them to build in; two holes are made in the partition to allow of free egress and regress, and the materials of which nests are usually composed, such as hay, elk's hair, down, feathers, and the ravellings of silk or cotton, should be put into a little net pouch, or bag, and hung from the roof of the cage near the perches. Some bird-fanciers recommend the washing of the breeding-cages with lime once or twice during the summer season, to keep the birds free from insects, but scrupulous attention to the cleanliness of the cage will always preserve its inmates from such annoyances.

LINNETS.

There are three varieties of linnets, the common linnet, familiar to every boy, the goldfinch, or thistlefinch, as it is also called on account of its feeding on thistle down, and the greenfinch or green linnet. But the common linnet is the singing bird, and many bird-fanciers say that the mixed breed of the canary and common linnet produces a sweeter singer than either of the birds unmixed. Linnets are easily tamed, show great fondness towards those who feed them, and seem to care less about being kept captive in cages than other birds. They also live a many years. The song of the common linnet is very sweet, very lively, and has many variations, and stands second to none of our small British song birds, if we except the blackcap, which is the sweetest singer of them all. As the common linnet is the most plentiful, readily procured, and sold cheap, it is the best song-bird the young fancier can start with, as it is not at all a difficult bird to rear. The male may be easily distinguished from the female by being browner on the back, having the first, second, third, and fourth feathers of the wings white up to the quill, and in the spring, by being crimson on the breast; the female, usually greyish on the back, streaked with dusky brown, and yellowish white, on the rump with greyish brown and reddish white, and on the breast these spots are tolerably plentiful; the wing coverts are dusky chestnut. When younglings in the nest, the males have a white collar, and some white tints about the tail and wings; and the females are generally more of a grey than a brown colour, and very much streaked on the breast. These birds are most usually taken in clap nets, and when secured, they should be put in store cages, and fed upon such seeds as you find they generally feed on, with the addition of a little bruised hemp-seed; the cage should be placed where the birds may not be molested for three or four days, after which time they should be taken out and put into separate cages, which are usually of very small dimensions and trifling value. These little  habitations are wired at the front and two sides, and the top and back are made of wood, painted on the outside green, and in the inside white; the receptacles for water and seed are commonly made of lead, but in superior kinds of cages a drawer for the seed, and a glass for the water, are often employed. The food most proper for these birds are the canary and summer rape-seeds (winter rape-seed is poisonous to them when in captivity, although not at all hurtful when they are wild), and a few corns of hemp-seed occasionally; seeded chick-weed, beet leaf, and lettuce-seed will be found beneficial if the birds be mopish; and if they are troubled with a looseness, a bit of chalk and some bruised hemp-seed, a stalk of plantain, and saffron in their water, are excellent remedies. If taken from the nest it may be taught to imitate the songs of the canary, woodlark, chaffinch, &c., and if kept by itself, to repeat tunes whistled to it. When they are taken so young, the food most recommended is moistened white bread, hard boiled egg, and soaked hemp-seed. Male linnets will pair with hen canaries, and their mule progeny can scarcely be recognised from grey canaries; their song is exceedingly beautiful, and they will learn tunes readily, and as we have before remarked, these mules are the best singers of the two. Linnets frequent hedges, bushes, and furze, and the skirts of woods; but as soon as autumn sets in, they take to the fields, and congregate in large flights; and in the winter they are wanderers, roving about in quest of food wherever the snow has not enshrouded the earth in its white robe, and are generally found at this season of the year near the sea-side. These birds have usually two broods in the year, and the young ones are sufficiently fledged in April to be taken. Their nest, which they take great pains to conceal, is often found in furze-bushes, the outside is formed of dry grass, roots, and moss, and the inside generally lined with hair and wool. The female lays four or five eggs, which are white, tinged with faint blue, and sprinkled with brown dots at the larger end. In flocking time the male linnet no longer shows the red on its breast.

THE BLACKCAP

We commenced with the common linnet on account of its cheapness, and the readiness with which it may be procured. We now give the crowned king of song, the sweetest singer in the summer band, not excepting, to our fancy, even the nightingale, and this is the blackcap. If you love real bird-music procure a blackcap at any cost, for he will make the whole house ring again, his song is so full, so sweet, so deep and loud, and so enriched with a variety of oily, silvery modulations, especially that long soft shake, which, though it sinks gradually into the lowest note a bird can utter, is heard as distinctly as the louder tones, and then just as you think it is about to die away, and you begin to anticipate the silence that must follow, higher and higher swells the song to the loftiest burst of melody, and you feel as if you wouldn't part with the bird for twenty times his weight in gold. When singing it distends its little throat, while the whole body quivers with delight, telling that it feels as much pleasure itself as it gives to the listener. Gilbert White, whose "Natural History of Selborne" every boy ought to read, for the sake of its beautiful descriptions of the habits of birds and animals, speaking of the blackcap, says its "note has such a wild sweetness that it always brings to my mind those lines in a song in Shakspeare's 'As You Like It'—

"And tune his merry note,
Unto the sweet bird's throat."

And I have no doubt in my own mind that Shakspeare was listening to the singing of the blackcap, or called to memory its notes, as he had often heard them when a boy in the green fields that spread around his native place, when he composed that beautiful and simple song which begins with—

"Under the greenwood tree,
Who loves to lie with me."

The back and wings of the blackcap are of an olive grey, throat and breast of a silvery grey, belly and vent white, sides of the head and back of the neck ash colour, and the top of its head black as night, whence its name. The female is a little larger than the male, and her distinctive marks are the cap brown, the upper part of her body reddish grey, inclining to olive, cheeks and throat light grey, breast, sides, and thighs light grey, tinged with olive, and her belly reddish white; she lays generally once a year, but sometimes twice; the nest is well built, and is commonly found in some low bush or shrub. The eggs, four or five in number, are of a pale reddish-brown, dashed with spots of the same hue, but darker. It is very fond of ivy-berries, and often builds in the ivy, when not too near the ground. Its favourite haunt is a garden or orchard, where, during the breedng season, it sings from morning to night. If you bring up the young ones, it is necessary to give them white bread soaked in milk, and if they are kept near other birds they will readily imitate their notes. If you are uncertain which birds are males,—for until the first moulting both sexes agree in plumage,—take a few brown feathers from the head, and their places will be supplied by black ones, if the birds are males; the song likewise will infallibly show the sex, as the males begin to sing as soon as they can feed themselves. Old birds are usually caught by nooses in the autumn, and they should then be fed upon elderberries and meal worms for a few days, so as gradually to bring them to their artificial food; this food should be bruised hemp-seed, and a paste made of bread soaked in water, and afterwards steeped in milk, with barley or wheat meal, or a paste made of hemp-seed, scalded and bruised, and white bread also soaked; these pastes should be mixed up fresh every morning, and when given to the birds some fresh raw lean meat chopped fine, should be added to them; the yolk of an egg boiled hard and crumbled into small pieces, is a very excellent variation to the general food; meal worms, ants' eggs, maggots of the bluebottle fly, &c. are also exceedingly good, and most vegetables are eaten by these birds with avidity.

THE BULFINCH.

Pick-a-bud, as the gardeners call this great destroyer of buds in spring, especially the young bloom of greengages, is a beautifully-marked bird, having a splendid red breast, a black head, and a pleasant looking ash-coloured back, which is varied by the black of his wing feathers and tail. He is very fond of singing while hidden amid dark fir-trees or thick impenetrable bushes, as if he liked to have it all to himself and not to be disturbed, and in such spots as these the nest is generally to be found, containing four or five eggs of a pale geenish white colour, dashed with dark orange-brown spots at the larger end. In a wild state the notes of the bulfinch are so low as only to be heard when very near the spot from where the bird is stationed; but there is something very sweet and plaintive in its low melodious notes, far more pleasant to our ears than that loud piping which they make after having been caged and taught. Though very few Naturalists agree with us in this opinion, they all admit that its notes are sweet in a wild state, but can only be brought out to perfection by teaching him to sing to the bird-organ. We contend that the notes are far sweeter when he is left to himself, and that after he is taught they are unnatural; there is a low silvery ring about the natural song of the bulfinch very pleasant to listen to in a room, but ten times sweeter when heard from some shadowy copse, when the winds that blow about you smell as if they had been out all day gathering perfumes from the May-blossoms.

The bulfinch possesses considerable powers of mimicry, learns to whistle airs with great correctness, and touches them off in so pleasing a manner, and with so soft a note, that it is often on this account one of the most highly-prized of cage birds. It may even be taught to repeat a few syllables distinctly, but its memory must not be taxed to remember too much. In Hesse and Fulda, in Germany, vast numbers of these little mocking birds are taught to whistle such airs as God save the Queen, the Hunter's chorus in Der Freischutz, &c.; they are principally brought over to England, where very high prices are frequently paid for them, especially if they are thoroughly accomplished.

In England bulfinches are not very plentiful, through a species of petty war having been carried on against them, from their destructiveness to wall fruits. They build twice in a year, and construct their unartificial nests in quickset hedges, or in retired parts of woods; the young are hatched in about a fortnight, and if you wish to rear the birds from the nest, take them when the tail feathers begin to make their appearance, and you may easily detect the young males from the females by their reddish breast. The food on which they will thrive best is rape-seed soaked in water, and mixed with white bread. When they can feed themselves, you may commence your course of tuition by whistling the airs you wish them to imitate, and you must not be discouraged by the length of time which may elapse before they can repeat the tune correctly. Soaked rape-seed, with the addition of a little hemp-seed now and then, by way of a treat, and some green food, such as chick-weed, water-cress, lettuce, &c., is the best food for these birds; sweets and other delicacies which some inexperienced persons recommend are highly injurious, and should on no account be administered freely. When moulting, a little saffron in their water, and a plentiful supply of green food, will be found very beneficial. If these birds are fed entirely on hemp-seed, they lose their variegated plumage, and become wholly black; indeed, the same alterations of colour, produced by feeding much upon that seed, have been observed on other small birds, such as the field-lark, wood-lark, &c. As these birds are not particularly restless, a middle-sized cage will do.

THE CHAFFINCH.

This is another very beautiful bird, whose singing generally is not much admired, though all admit that the bird itself is a beauty to look at. For our part, we consider it a very pleasant singer, not first-rate certainly, but then every bird is not gifted with the voice of the blackcap, and for our part we like variety in the songs of birds, and this is what makes listening to them in the fields and woods so delightful. There is something very curious in the female chaffinches quitting this country about November, and leaving all the males in a melancholy state of bachelorship, or temporary widowship, if we may be allowed so to call it, behind. Then the nest of the chaffinch, for neatness and beauty of construction, beats all the nests you ever saw, and so closely resembles the foliage and branches amid which it is built, that only a keen and practised eye would be able to discover the nest. The outside is covered so beautifully with moss and lichen—you know what we mean by lichen, the flakes of gold and silver colours, which are a kind of fringe that is often seen on the stems of trees or growing in damp places; then the nest is so neatly composed of hair, wool, and feathers, that were you to try for a hundred years, with all the materials at your hand, and all the tools for use that could be invented, you would not at the end of that time be able to form anything half so neat and beautiful. The eggs, too, which seldom exceed five nor less than four, are of a pretty reddish-brown colour, marked with dark spots at the large end.

Chaffinches build in hedge-rows, and the young ones are hatched about the beginning of May; they may be taken when about twelve or fourteen days old, and should then be fed upon the crumb of white bread and rape-seed soaked in water. The males may be distinguished from the females, even at that early age, through the breast being more tinged with red, the wings blacker, and the lines crossing them whiter, and from the circle round the eyes being of a deeper yellow colour; if you are uncertain respecting the sex, pluck out some of the breast-feathers (they will be renewed in a fortnight), and if the red tint is visible, you may be certain that it is a male bird; if otherwise, a female. The plumage of a full-grown male chaffinch is extremely beautiful; its forehead is black, beak blue in the spring, but after moulting and during the winter, white; the crown of the head, and the hinder part and sides of the neck, bluish ash colour; the sides of the head, the throat, fore parts of the neck, and the breast, are of a vinous red; belly, thighs, and vent white, lightly tinged with red; the back is reddish brown, changing to green on the rump; the greater and lesser coverts of the wings are tipped with white, the bastard wing and quill feathers are black, edged with yellow; tail black, the outermost feather edged with white, and legs dusky colour. The female differs considerably from him, her head, neck, and upper part of her back, are greyish brown; the under parts of her body dusky white, slightly tinged with reddish grey on the breast. Chaffinches frequent copses, orchards, and forests, and old birds and branchers may be taken with clap-nets in June or July. Rape-seed is that on which they will thrive best, and to which a few corns of hemp-seed may be sometimes added to incite them to sing; chickweed and other green food they also like. In Germany, chaffinches are so highly valued that very high prices are given for them if they possess a fine song; a common workman will give sixteen shillings for a bird whose notes he considers good, and will frequently live upon bread and water until he can save money to purchase the desired object.

THE GOLDFINCH.

This beautifully-marked bird is a favourite with everybody. It is such a pretty thing to look at, and has so many graceful attitudes when jumping about in its cage; or, as we once heard a plain countrywoman say, who was a great lover of birds, "It has such a many winning ways with it, that one can't help liking it, if even we try." Then, to say nothing of its singing, although that is very pleasant to listen to—a little deficient in variety, perhaps, not so quick in picking up tunes as a few other birds; still it can do no end of things which better singers cannot do; and in a wild state its song, which may then be heard at almost every season of the year—not excepting even winter, when the weather is mild—has a brisk, cheerful, heart-stirring ring about it. Its nest, too, is very pretty; all kinds of soft substances are felted together, just as a hatter would make a hat, not a single particle projects—wool and feathers, hair, moss, lichen, and bark, are all beautifully blended together, and in such a way as no machinery could do it, without bruising and spoiling the materials. The eggs are small, of a delicate whitish tint, and beautifully spotted over with orange brown. Its forehead and chin are of a beautiful scarlet colour; its bill is white, tipped with black, and a black line passes from each corner of the bill to the eyes, which are dark; the top of its head is black, and the same colour extends downwards from the nape on each side, so as to divide its cheeks, which are white, from a spot of white on the back of the neck; its back and rump are of a cinnamon brown tint, sides the same, but rather paler; belly white; greater wing coverts black; quills black, barred in the middle with yellow, and tipped with white; tail feathers black, with a white spot on each near the end, and legs of a pale flesh colour. It will learn to come and go at command, and to perform a variety of tricks, such as firing a cannon, counterfeiting death, letting off a cracker, or pulling up a bucket of water. In a wild state, goldfinches frequent brambles and thickets, and woody districts, which are interspersed with fields; they are also partial to thistle fields, where they congregate in large flocks; they are generally taken in clap-nets. In the summer they are tender, and rather difficult to rear, but in winter they will soon sing after their capture. When taken young, they should be fed upon white bread and milk, with a little of the flour of ground canary-seed, for five or six weeks, and then give a little canary-seed in addition, and the sooner you can bring them to the canary-seed alone, the better. When full grown, they may be fed upon poppy and canary-seeds, with lettuce and rape-seeds occasionally by way of a change; green food, such as chickweed, watercresses, lettuce, and endive, should also be given sometimes. If your bird is troubled with a looseness, a little bit of chalk should be put in the cage, and red sand strewed over the bottom of it. A square cage is the best shaped one for this bird, as it does not admire hopping about the upper part. Goldfinches will pair with canaries, and produce fruitful mules; it is better to pair a male goldfinch with a female canary, than the reverse; in plumage the mule birds are very pretty, blending the richness of colour of the goldfinch with the yellow of the canary, and in point of song they are exquisite.

THE REDPOLL.

This bird as a singer is nought, though it is a great favourite with most boys, on account of its tameness and the readiness it can learn almost anything, except singing. The upper part of its body is a dark

brown, and the feathers are tipped with a paler tint of the same colour; the feathers of its neck and breast are rose-coloured, edged with white; the rump is rose-coloured also, and the rest of the under part of the body white; the greater and lesser wing-coverts are bordered with dirty white, which forms two light bars across the wings; the forehead is of a most brilliant crimson; bill light brown colour, dusky at the point; legs dusky. The female is not so strongly coloured, neither has she any rose-tint on her breast; the upper part of her body is speckled with brown and white, and her breast is slightly spotted with the same hues. These docile birds may be taught to draw up a little bucket of water, like the goldfinch, to come and go at command, to hop along a species of ladder, composed of small wooden pegs driven into a wall, about six inches apart, and so arranged as to form a gradual ascent and descent, as

shown in the annexed figure, and also to hop from one thumb to another, held at some distance asunder. Before commencing their tuition, they must be deprived of the power of flying away, either by clipping the feathers of one wing, or by pulling out some of the flight feathers altogether; the latter method, though more cruel, is too frequently preferred, though by this plan the wing is more quickly restored to its natural condition again, the feathers being renewed in about six weeks, while by the other plan they are not restored until the next moulting time. By the time the wing is properly grown, the birds generally become so tame as to render a repetition of the operation unnecessary; indeed, so thoroughly domesticated will they become, that they may be allowed almost perfect liberty; they may even be taken out to the distance of half a mile, or a mile from home, and they will return, although sometimes after a week's absence. The food proper for them is canary, rape, and flax seed, mixed; sometimes also a few grains of hemp, as a treat, with maw-seed, or a little saffron in the water, as a medicine.

THE REDSTART.

This is another handsome bird, and a very pleasing singer. It derives its name from the bright mahogany-red colour of its upper tail coverts and tail feathers, which show like dashes of fire as the bird flits or starts from one bough to another, when alarmed. It is fond of frequenting the borders of woods, or wherever a clump of trees or a copse offers shelter; and its nest is very often found in old walls covered with ivy, in which it lays five eggs of a greenish blue, and rather difficult to distinguish from those of the hedge-sparrow. It often sings as it flies, and is constantly changing its position; and, although the notes are not very loud, they are remarkably sweet. Gilbert White says, it will sit on the top of a tall tree, and sing from morning to night; it avoids solitudes, and loves to build in orchards and about houses. It is, like most warblers, a migrating bird, and visits England at the latter end  of March, or the beginning of April, and disappears about the last week in September. Its plumage, as before remarked, is very pretty, for its cheeks, throat, fore-part, and sides of the neck, to just above the eyes, are black; the crown of its head, hinder part of the neck, and back, deep blue grey; breast, rump, and sides of a fine glowing red, inclining to orange, and this colour extends to all the tail feathers, excepting the middle ones, which are brown, and the wing coverts; its belly is white, and so is its forehead; its bill, eyes, feet, and claws are black. The female differs considerably from the male, as her colours are not so vivid, the top of her head and back being of a grey ash colour, and throat white. The young ones, if it is wished to rear them from the nest, to tutor them to warble airs, or in order that they may improve their own native strains by imitating the songs of other birds, should be taken when the tail feathers begin to grow, and fed upon ants' eggs and bread soaked in water. When wild, redstarts feed upon insects and berries; and when in captivity they must always be supplied with insects; they are particularly partial to ants' eggs, meal worms, and common maggots, and in general such food as is given to nightingales will be found best suited to them.

THE REDBREAST.

Robin Redbreast is a beautiful bird, remains with us all the year round, hail, rain, frost, or snow, is as great a favourite with the old grandsire of threescore and ten as he is with the little thing that can just toddle, and who stumbles and drops the pinch of salt which it believes would capture Robin if only dropped on his tail. For our part we do not like to see him a prisoner in a cage; he comes so near  to our homes that we can hear his song almost at any time if there is only a morsel of garden ground big enough for him to alight within, and it is like breaking faith in the trust he has in man to make him captive. It is different with those that were taken when young, they do not so much miss their freedom; and those that were hatched in spring will sing in autumn although they have but just moulted, and for the first time began to show the red feathers on their breasts. For our part we never in our boyish days took the nest of the robin, though we were great birds'-nesters, and have found it scores of times by the sides of banks, under the roots of trees, or amid masses of ivy that grew low down, we always left those beautiful pale grey red-spotted eggs untouched for the love of the good old ballad of the "Babes in the wood" whom Robin Redbreast "painfully" did bury beneath the leaves :—

> "Leaves of all hues, gold, red, and green,
> Ruins of summer bowers;
> A thousand times more beautiful
> Than all her choicest flowers."

Gilbert White says, "I knew a tame redbreast in a cage that always sung as long as candles were in the room; but in their wild state no one supposes they sing by night." They do, and we have often heard them singing during a warm moonlight night. The female breeds twice a year, laying each time from four to six or even eight eggs; the young birds do not show any beautiful colours until they have moulted. They may be reared from the nest without difficulty, if white bread soaked in milk is allowed them. The plumage of the redbreast is pleasing though not showy, the head and all the upper parts of its body are brown, tinged with a greenish olive; its neck and breast of a fine deep reddish orange tint, and a patch of the same colour marks its forehead; its belly and vent are of a dull white; its bill is slender, eyes full, black, and expressive, and its legs dusky. The female is somewhat smaller than the male, and the reddish tints on the forehead and breast are not nearly so brilliant in colour. These birds may be taken in the autumn in nooses baited with elderberries, on which they then very much subsist, and which berries also should be given to them when they are first put into confinement, as well as earth and meal worms, &c.; they will soon become accustomed to almost any kind of food, such as cheese, crumbs of bread, and little pieces of meat, and their proper diet is the same as that of the nightingale. The redbreast is generally kept in a cage like the nightingale, and a little pan of fresh water should be frequently put in, so that the bird may enjoy a bathe, a diversion of which it is very fond. It may be taught to come and go at command, and may be so tamed as to eat from the hand.

Boys must have birds to rear and feed, as sportsmen must have game to shoot, and if the redbreasts are taken in the nest, it had better be while they are too young to pine after the hen, and that is about a fortnight after they are hatched. They must be kept warm but not hot; when a little sheep's heart, minced very small, may be given them in small quantities at a time, and often. Never try to make a young bird gape for food, it will open its beak readily enough if in want of it. Remember all young birds must be kept very clean; and that a small quill cut open and round at the end is the best thing to feed them out of. Too much food prepared at a time soon loses its sweetness; a bird likes everything it eats fresh and sweet.

THE SKYLARK.

There are three kinds of larks, all of which are easily recognised as belonging to the species, by their long hind toes. They all, like the skylark, sing while on the wing; in addition to which the titlark or meadow-pipet sings while sitting on the ground, and the woodlark while perched on the branch of some tree. But the skylark is the sweetest songster of the three. He is the bird that Shakspeare fancied went singing up to the very gates of heaven; the minstrel of the sky, who makes all the gold and silver pillars in cloudland echo when he warbles in his great star-roofed skyey hall. This is the bird that sleeps beside the daisies and among the gentle lambs; that makes its nest in any hole that it finds in the ground; the print of a horse's or bullock's hoof serving as well as anything else in which to deposit its five greenish-white brown spotted eggs. Tens of thousands of these sweet sky-singers are caught, sent to market, sold, cooked, and eaten every year; and if that isn't worse than keeping them in a cage to sing, why the deuce is in it. How differently we treated them in our boyish days, when we hunted for the green sod which had the most daisies upon it to place in our lark cages for the birds to stand and sing upon. We should as soon have thought of cooking and eating a "blessed babby," as that little singing angel of the heavens, the skylark. But great guttling fellows can't care for the singing of a bird which they would rather eat. They would have fish, flesh, and fowl for dinner, if they broiled the gold-fish that was swimming in the glass globe, roasted little Jack's pet guinea-pig, and baked the canary under a crust, though it lifted up its pretty little toes imploringly and said "Pray, don't." They care nothing at all about its praises having been sung by poets of all grades, good, bad, and indifferent, nor its pleasing note being the delight of every lover of nature. The skylark will readily imitate the songs of other birds, and also learn tunes; in confinement it sings during half the year, and may be tamed so as to come and eat from the hand. The skylark breeds twice a year in temperate seasons, and forms its nest on the ground in high grass, or a wheat-field, or on a common or heath; the young are hatched by the end of April, and may be taken when about ten or twelve days old; they should be put in small cages and fed with poppy seeds and the crumb of white bread soaked in milk; to this food, ants' eggs and a little lean meat will prove a nourishing addition; some persons give their birds bruised hemp-seed mixed with hard-boiled eggs chopped fine. When young, the male birds may be detected from the females by being yellower in colour, and when arrived at maturity they are larger in size, and not so white on the breast, neither have they so many black spots on the back and breast. The food for full-grown birds consists of German paste, poppy and bruised hemp-seeds, bruised oats, crumb of bread, and an abundant supply of greens, such as water-cresses, chickweed, lettuce, &c., a small quantity of minced lean meat and a few ants' eggs may be added sometimes for a treat. If the bird is unwell, or becomes loose, grate a little cheese into his food, and give him some wood-lice three or four times a day; a blade or two of saffron, and some liquorice in his water, will also be of service, as likewise a spider occasionally. Larks are caught with clap-nets, and also with nooses, and in dark nights with the trammel.

The lark's cage is always plain in its appearance, being painted green without and white within. The roof is gabled, the back boarded, and it should have a drawer at the bottom for convenience in cleaning; the front of the cage, from about an inch and a half from the bottom to the lower part of the gabled roof, is bowed, and on the floor of this bow a sod of clover or grass is placed, and this sod should be renewed every other day; the sides of the cage are wired, and the places for seed and water may be outside; in some cages a drawer for the seed is substituted for the little box or glass. Plenty of sand should be spread over the bottom of the cage. Larks are the only birds which sing while winging their way up into mid-air, and their clear strains may be heard even when they have soared far beyond the reach of sight.

THE TITLARK.

The meadow-pipet (for that is its real name), like the skylark, builds its nest on the ground, where it is generally found under a tuft of tall, close-grown grass, and in which it lays from four to six eggs of a light-brownish colour spotted with the same hue, though of a darker tint. It is the smallest of the lark tribe, and is a handsome and slender bird. In the arrangement of its colours it resembles the skylark, but is of a rather darker and more greenish brown than that bird. Its breast is elegantly marked with black spots on a light yellow ground, belly light ash colour, faintly tinged with dusky streaks; its tail is almost black, the two outer feathers edged with white; its legs are yellowish, feet and claws brown, its bill brown, tipped with black, and its eyes hazel. The young may be reared from the nest, if fed upon ants' eggs, and bread soaked in boiled milk, and a few poppy seeds; they learn to imitate the songs of other birds, but never arrive at great perfection in their imitations. The time for catching old birds, or branchers, is from the end of March to the middle of April, for if taken at a later period, they will not sing much during the first summer; clap-nets are the means generally employed for the purpose, but limed twigs are also sometimes used; to take them by the latter method, it is necessary to have a caged titlark as a call-bird, which you take out with you, and when you have discovered a wild one, put your call-bird down on the ground, at a few yards from where you heard the other, and scatter a few well-limed twigs round the cage; secrete yourself, and the wild lark hearing your caged bird, will approach, and most probably settle on one of the twigs; the instant he perches, you must rush forward and take him, or he will free himself from the twigs, and escape; tie his wings, and when you put him in a cage, supply him with some meal-worms, ants' eggs, or caterpillars, and bruised hemp-seed, and accustom him, by degrees, to skylark's food—that is, give him meal-worms and ants' eggs plentifully for the first day or two, then mix a few poppy and hemp-seeds with the worms, and increase the quantity of seeds gradually. As this lark perches, its cage must have two bars, but in all other respects it should be made like the skylark's.

THE WOODLARK.

If the woodlark cannot soar so high nor sing so sweetly as the skylark, it can do what the latter can't, and that is, it can either sit and sing on the branch of a tree, or have a fly and carry its music along with it, just as it pleases. But though so partial to perching on a branch, like the sky-lark, it builds on the ground. Some bird-fanciers say it possesses a more musical and sonorous note than most other singing-birds, but its imitative faculties are not very good, for unless it is reared from the nest near some other birds, it will not learn their strains. In plumage it resembles the titlark, but the upper parts are not so clearly defined; a white stripe passes from the bill over each eye, towards the nape of the neck; its under parts are white tinged with yellow on the throat, and red on the breast or spotted with black; its tail is not so long as that of the other larks, consequently the bird looks thicker in its shape. It builds a tolerable nest among heath, in hedges, high grass, and under little hillocks. The young birds may be reared from the nest upon bread soaked in milk, and ants' eggs. In June and July, woodlarks may be caught with a clap-net, and their haunts are principally pasture-lands, gravel-pits, and heaths. The best food to give them after their capture is a mixture of poppy-seeds, oats, young wheat, fresh and dried ants' eggs and meal-worms, minced sheep's heart, mutton, veal, or lamb. Some persons, instead of the above food, give their larks finely-bruised hemp-seed mixed with bread, and some ants' eggs, twice or thrice a day, and a piece of bread which has been soaked in milk. The bottom of the cage should be covered with red sand, and that and the water changed every day. When the bird is out of order, give him a few hog-lice every day; if he is troubled with a looseness, put some mould full of ants' eggs at the bottom of his cage, and grate a small quantity of Cheshire cheese or chalk amongst his victuals; a blade of saffron and a little piece of stick-liquorice in his water will be of good service in clearing his voice and causing him to sing freely and powerfully. The cage may have perches, as the woodlark does not always roost on the ground.

THE BLACKBIRD.

This is the "ouzel-cock with golden bill" so often mentioned in our old ballad poetry, and pleasant it is to hear him often as early as February, reminding us with his song that spring is nearer than it was, the days longer, and that the pretty primroses in warm sheltered places are beginning to show their pale golden-coloured flowers. During the first year, the males bear so close a resemblance to the females, that those only who are well versed in "birdship" can distinguish the difference, as it is not until the second year that the male shows his golden bill. The blackbird is a dusky gentleman of rather solitary habits, and excepting the society of his dark lady, cares very little for company, but is very much given to musing and singing in gloomy woods and close-woven thickets through which some stream flows, for it must be near water. In its wild state it feeds on berries, fruits, insects, and worms, and hardly anything seems to come amiss to it. He has been heard to imitate part of the song of the nightingale and even the crowing of the farm-yard cock, and seeming to enjoy the fun. Blackbirds pair early in the spring, and the first young are often hatched by the end of March; they have usually two or three broods a year, and lay from four to six eggs of a greenish colour, spotted and streaked with brown. When the young are hatched, the males are said to be blacker in tint than the females, and therefore, as some suppose, very easily distinguished. If you wish to teach the birds to whistle airs, you must take them when the quill feathers are just beginning to appear, for they are easily reared on white bread soaked in milk, a little lean raw beef, and a few worms dipped in water. The most suitable food for these birds, when mature, is bread, meat, either raw, boiled, or roasted, and woodlark's food, with the addition of a little bit of apple sometimes.

The cage for blackbirds should be large and roomy; the old-fashioned peak-topped wicker cage is so generally known, that it is useless giving a representation of it; but the annexed figure shows an improved plan for a cage, in which the rough, homely wicker is blended with polished mahogany; the top of the cage is gabled, and that and the back are of wood, the two sides and front have wicker rails in place of wires, which are strengthened by passing through horizontal bars of wood; the water and food are put into the two little boxes at the sides, and there is a drawer in the bottom for the greater cleanliness; a little pan of water should be often put into the cage, as this bird is very fond of amusing itself in a bath. If the bird is unwell, a large spider and a few wood-lice will be a good diet for it; and a small quantity of cochineal in the water will also prove very serviceable, and make him gay and brisk. Hog lice are also considered excellent restoratives, but they should be administered with discretion, lest the bird's appetite for other food is taken away by having a superabundance of such (to its taste) delicious fare. The natural note of this bird is pleasing, but it sounds better in the open air than in a room, as there are many noisy tones intermingled with the others, which interrupt the flowing character of the melody. Blackbirds may be taught to whistle tunes and repeat short sentences, in the same manner as the bulfinch; indeed, they are preferred by many persons to that bird, as their acquired note is particularly musical. They are never kept in aviaries, for when shut up with other birds, they plague and harass them incessantly; there may be exceptions, but this is the case generally. The blackbird never forsakes us, but stays with us all the year round, and we think that if only for filling the air with such sweet music as he does, while even winter still reigns, often in all his gloom and chilliness, the sportsmen ought to spare pretty "golden bill." A man who would shoot at a blackbird while it is singing would, we think, hardly refrain from pointing his gun at a little lonely child in a solitary green lane.

THE THRUSH.

The thrush, or throstle, is the "sweet mavis of our old ballad lore," and under the latter name is often alluded to by our early poets, who tell us what pleasure they felt in wandering through the "greenwood shaw to hear the merle and mavis sing." It even commences singing earlier than the blackbird, and when the weather is a little mild may often be heard in January. It frequently begins to build before there is a leaf out on either hedge or bush, and, as if aware that its young when hatched may have a good deal of cold weather to contend against, it plasters its nest with a thick coating of mud, which when dry defies the wind to find an entrance. It is a very handsome bird to look at, with its beautifully speckled breast, to say nothing of the rich fulness and sweetness of its song, which may be heard throughout the summer, and particularly at the grey dawn of morning and in the still evening twilight. Its bill is dusky; eyes hazel; the lesser coverts of its wings and its back and head of a deep olive brown; the tips of its wings white; the lower part of the back and rump tinged with yellow, and its cheeks of a yellowish white with brown spots; and on the breast and belly larger spots of the same colour likewise, on a yellowish white ground; the tail feathers brown, the two outermost tipped with white; legs yellow, and claws black. The female is very similar to the male in his plumage, save that she is not so brilliant in colour; she lays five eggs of a bluish-green colour, which are generally, though not always, spotted with a deep reddish-brown. The first brood is mostly ready to fly by the end of April. If the young are taken from the nest when half-grown they may be easily reared on white bread soaked in milk, and taught to whistle airs. When mature, they are to be fed with the same food as the blackbird, and if out of order treated in the same manner; they require a plentiful supply of fresh water, both for drinking and also bathing, to which they are extremely partial. Both males and females will begin to record as soon as they can feed themselves; the males will get on the perch and utter their notes in a low key, while the females will perform theirs by jerks. If you are not quite positive as to the sex of the birds, keep them till after moulting, when the males will start into full song. The thrush's cage should be large and roomy, as it is a very animated bird, and brisk in all its movements.

THE CANARY.

These delightful little warblers are not natives of Europe, having been originally brought from the Canary islands; they have, however, become in some measure acclimatized, and are probably more esteemed by all classes of people than any other species of song bird. In a wild state their colours vary exceedingly, some being grey, others white, some chestnut, some yellow, and others blackish; and it is by an intermixture of these colours that the varieties now in fashion take their origin. The yellow or white-bodied birds are the most esteemed when the wings, tail, and head (especially if crested) are yellowish dun; the next valued are those of a beautiful rich yellow, with the head, wings, and tail greyish; grey birds with a yellow head and collar, and yellow with a greenish tuft, are also much admired. It is difficult to distinguish the male bird from the female, but as a general rule it may be observed that he is rather larger and longer in the body, more elegant in his form, and higher in his shanks than the female; he is also longer from the legs to the vent, and particularly taper in that part, and if you blow the feathers up, his vent appears larger, and the orifice smaller, than in the female. Another test for distinguishing the sexes is their colour, the male being brighter than the female, especially round his eye, where the colour is a deeper yellow than any other part of his body. Those birds which introduce amongst their own notes some of the nightingale and woodlark's songs are the most esteemed, and it is highly necessary when purchasing a bird to hear it sing before you complete the bargain, as many females, particularly old ones, by uttering a few unconnected notes, have been mistaken by unskilled persons for males. Some birds not only imitate airs with correctness, but even learn to pronounce distinctly a few short words. The canary breeds four or five times a year, and lays four, five, and sometimes six eggs each time. The birds should not be paired till the middle of April, and they should be put either in a very large cage made for the purpose, or else allowed to range about a room. If you put them in a cage, let it be so large that the birds may have room enough to fly about with freedom. It is a good plan to have two little boxes for the birds to build in, as they are apt to go to nest again before the young ones fly. Birds which are to be paired for the first time ought to be placed in the same cage for a few days, that they may become accustomed to each other. If you give the birds the range of a room, nest boxes should be nailed up in various corners, and moss thrown about the floor; if a wire-gauze blind can be fastened across the window, so that the latter may be occasionally left open to allow fresh air to blow freely into the room, it will add materially to the health of its inmates. You must take care to furnish your birds, whether in the cage or room, with some fine hay, horse hair, hair of cows and elks, and hogs' bristles, in order that they may make their nests. When the hen has laid about six eggs she prepares for the process of incubation, which usually lasts thirteen days, and when the young are hatched it is necessary to put a little jar by the side of the feeding trough, containing some hard-boiled egg chopped very fine, and a small piece of white bread which has been steeped in water, and afterwards squeezed almost dry; in another vessel some rape-seed which has been scalded, and then steeped in fresh water, should be put, and the greatest care must be taken that the rape-seed is not sour, else it will certainly kill the young ones. When you bring up the nurslings by hand, the utmost attention must be paid to them, and the food most appropriate is a kind of paste made of white bread, bruised rape-seed, and a little yolk of egg tempered with water. This paste must be given to the little ones on a thin small piece of wood, shaped like a spoon, and they should be fed twelve or fourteen times a day, every time giving them about four beakfuls. The young ones must be suffered to remain with their mothers for about twelve days, by the end of which time they will be fledged, and on the thirteenth day they usually begin to peck up the food for themselves; they will require to be fed by hand for twenty-three or twenty-four days, and at the expiration of that time they may be put into separate cages, the bottoms of which should be strown with fine hay or well dried moss. They must, however, be fed for some weeks on the before-mentioned paste, with the addition of the general food of a full-grown bird, and as they gain strength and vigour the paste may be gradually withdrawn, until at length they become accustomed to their ordinary food, which should consist of summer rape, canary, and poppy, and bruised hemp-seeds, with oatmeal and millet occasionally in the summer as delicacies. Green food, such as chickweed, groundsel, radish, lettuce, water-cresses, plantain, &c., should on no account be omitted, neither should a daily supply of fresh water for bathing in be forgotten. The cages for canaries are more showy and elegant in their shapes and materials than those for any other birds, gothic, Chinese, and arched being amongst the most usual patterns; and within the last three or four years very pretty dome-topped cages, made of brass wire, with surrounding bands and stands

of brass, have become very fashionable. The gothic cages of wood are made of mahogany; the tops and sides wired, as is also the front in a fanciful manner, and they are usually fitted up with two perches, water and seed glasses, and sliding drawers. The brass dome-topped cages are likewise fitted up with perches and sliding drawer, but instead of glasses for seed and water little japanned cups are fastened to the lower perches. Canaries being rather tender, they should never be kept in a cold room during winter, for such attempts at naturalization are highly prejudicial to the poor birds, and will in all probability cause their death. In summer they may be hung at an open window where they can enjoy the bright sunshine, and while revelling in its brilliancy and warmth, they pour out their gushing melodies with renewed vigour.

THE NIGHTINGALE.

"When unadorned adorned the most," may be applied to the sweetest singer upon earth, for the plumage of a nightingale is about as plain as that of a common house-sparrow. It arrives in England about the middle of April, but when it leaves our shores is somewhat uncertain, though it has rarely been seen after the closing in of summer—never, we believe, in autumn, unless the bird had met with some accident, and was unable to fly. The female commences her nest early in May, and, like the skylark, always builds on the ground; the eggs, four or five in number, are of an olive-brown colour. The young ones leave their nest before they are well able to fly, and amuse and strengthen themselves by hopping from branch to branch. The nightingale, in its natural state, is so wrapt up in its singing, that a stone thrown at the bush in which it is stationed, unless it happens to hit the bird, will not stop its song. Scare it away by all the noise you can make, and you will hear it again, a minute or so after, singing in the new spot it has chosen. It is often taken in traps baited with meal worms, which it is necessary to set near the spot where they have been heard to sing; yet they are so unsuspecting in their natures, that they will notice the fixing of the snare, and then fall into it. The retreats in which these birds mostly delight are woods, groves, coppices, quickset hedges, and thick brambles, wherever the air is not too cold. When you have secured one, tie his wings together with a little piece of thread, and give him some ants' eggs and meal worms. The period of incubation is generally about a fortnight, and in plumage the young birds, before the first moulting, bear so little resemblance to the parent birds, that they might almost be taken for a distinct species; the upper part of their bodies being of a reddish grey colour, and yellowish white spots ornamenting the head and the wing coverts, while the under parts of their bodies are of a rusty yellow tint, with brown spots on the breast. The males may be distinguished in the nest, as being marked with white, and by having white throats; the females are redder and browner in colour than the others. Young females sing as well as males for a month or so, but in a weaker and more interrupted style. When you take the young from the nest, they must be fed with ants' eggs, mixed with soaked white bread; and ants' eggs are plentiful at this season. When the bird is full grown, the whole of the upper parts of its body are of a rusty-brown colour, tinged with olive; the under parts pale ash colour, verging to white at the throat and vent, the quills are brown with reddish margins, its bill is brown, eyes hazel, and legs pale brown. The female is very similar to the male. The sides and back of a nightingale's cage are made of wood, and the front only wired; the roof may be gabled, and an inch or two below it a ceiling of baize, or some other soft material, must be strained, so that the bird may not hurt itself as it rises upward when singing—a peculiar habit—and the perch must be padded for the same reason; just below the bottoms of the wires in the front of the cage another and smaller perch is put, supported upon two stems. The cups for food and water are placed in holes made in two small shelves, which are fastened in the front corners of the cage. The bottom should be furnished with a sliding drawer, and the door is usually made at the back of the cage. The cages are generally of mahogany, and an attempt is made at architectural decoration in the front, from its being embellished with a pediment, and  and an urn-shaped ornament. A little pan of water should be put into the cage for the bird to wash himself in, and it is highly necessary to keep the cage perfectly clean, and in a room, the temperature of which is never below temperate, as the nightingale is extremely susceptible of cold. If the bird is out of order, if it puffs up its feathers, shuts its eyes, and sits for hours with its head thrust under its wings, ants' eggs, spiders, and saffron in the water are the best remedies; if the dung is rather looser than ordinary, a little hempseed ground fine, mixed up with minced sheep's heart and egg, must be administered. During moulting this bird requires succulent food, and a spider now and then by way of a drastic.

DISORDERS OF CAGED BIRDS.

Although we have appended to the description of each bird a brief account of a few of the disorders they are subject to, we shall now notice some of their complaints separately, so as to enforce them more strongly upon the mind of the young bird-fancier. We also give a list of the usual remedies that are applied, not only from our own experience, but from the highest acknowledged authorities.

HUSK, or ASTHMA, is a disease of not unfrequent occurrence amongst caged birds; it sometimes arises from cold, proceeding from neglect, and sometimes from the birds having had too much hemp-seed, which, although all birds like it, is very injurious, as it is overheating, and incites them to gorge. The curatives are aperients, such as a spider or two every day, and endive and water-cresses; occasionally boiled bread and milk, and bread soaked in water, are very good. Some persons recommend a drink, made by boiling linseed and stick-liquorice in water, as being very excellent. The symptoms are, shortness of breath, and frequent opening of the beak, and if alarmed, keeping it open for some time.

The PIP is a cold which stops up the nostrils, and hardens and inflames the membrane which covers the tongue. The symptoms are opening of the beak, its yellowness at the base, and the dryness of the tongue. A composition of pepper, fresh butter, and garlic is the best remedy, and a feather must be passed up the nostrils, for the purpose of opening them. In large birds, such as domestic fowls, it is usual to remove the inflamed skin, by tearing it off with the nail.

The SURFEIT is a disorder to which young birds are particularly subject, arising either from giving them too much food, or from their own gluttonous propensities, when they feed upon the same kind of diet. The symptom of this disorder is a swelling under the belly, owing to the bowels sinking down to the lower part of the body, and sometimes turning black. The same kind of protuberance often shows itself when the bird is suffering from a cold, and the disorder is then termed a swelling; in this case the swelling is at first white, but it afterwards turns red, as in the surfeit. The utmost care must be taken with the poor little sufferers, as few survive the last stages of this malady. Some fanciers recommend whole oatmeal as a good cleansing food during the first three or four days, putting saffron in the water at the same time; if, however, the bird is too loose, maw-seed and bruised hemp-seed, with some groundsel, and saffron in the water, should be substituted. Boiled bread and milk with maw-seed put in it is by some reckoned good, as are also millet, hemp, canary, and rape seeds boiled together, with some hard boiled egg minced very small, and about as much lettuce-seed as any of the other kinds added. Treacle may be put in the water which you give the birds, before furnishing them with their daily supply of seed.

SWEATING is a disorder to which some hen canaries are subject during the time of incubation, or while they are nursing their young.

To stop this complaint, which will, unless checked, kill the young brood, some fanciers advise the hen to be washed in salt and water, then dipped in fresh water to neutralize the effect of the salt, and afterwards dried as quickly as possible, either in the sun, or with the help of dry soft cloths before the fire. This bathing and drying should be repeated once or twice a day, until the little patient recovers. The best plan to cure this disorder, however, is to take the hen away, and keep her from sitting.

OBSTRUCTION OF THE RUMP GLAND.—This gland furnishes the oil with which the birds trim their feathers; it sometimes hardens and becomes inflamed, and unless the sufferer pierces it himself, it must be cut or pierced with a needle, the inflammatory matter pressed out, and a little fine sugar dropped on the place; this simple remedy often effects a speedy cure.

LICE.—The insects by which many caged birds are annoyed, are principally produced from their own slovenliness. Old wooden cages are very liable to be infested with these pests; for the insects being very minute, they get into the smallest crevices, and remain housed during the day, making their appearance only at night. Old cages should therefore either not be used, or else very frequently attended to; and if a pan of fresh water is put into the cage, it will be of great service in promoting the cleanliness of the birds, as it will enable them to sprinkle themselves.

OVERGROWN CLAWS AND BEAK.—When a bird's claws grow long, it is necessary to cut them, otherwise they are very inconvenient; they must not, however, be cut so short as to draw blood, else the bird will be lamed. The beak also requires paring sometimes, and the scissors for this purpose, and for the claws, should be perfectly sharp.

MOULTING.—While suffering from this malady, the birds must be taken great care of, supplied with plenty of nourishing food, and kept warm. Millet, lettuce, canary, maw, and hemp-seeds, bread soaked in water, and green food should be given to those birds which subsist upon seeds; and an additional supply of meal worms and ants' eggs to those which feed upon insects. A little saffron, or a rusty nail, may be put into their water with advantage.

LOSS OF VOICE.—Male canaries sometimes suffer the loss of voice after moulting; they should then be supplied with a paste composed of bread pounded very fine, mixed with well-bruised lettuce, and rape-seeds, tempered with a little yolk of egg and water.

COSTIVENESS may be removed by giving such aperients as spiders, plenty of green food, and boiled bread and milk; to those birds which subsist upon meal worms, one bruised in sweet oil and saffron will be an exceedingly good alterative.

CONCLUDING OBSERVATIONS.

If the young fancier wishes to preserve his little captives in health and song, he must be scrupulously attentive to the cages in which they are kept; they should be cleaned out twice or thrice a week, and the perches scraped once a week at furthest, for negligence in these particulars engenders many evils which the birds only can suffer, such as gouty feet, loss of claws, &c., besides the inconveniences unpleasant to the fancier, arising from the scent of sick birds.

Some fanciers recommend the use of a lime wash for the inside of the breeding cage, once or twice during the summer months, but if careful cleaning will not keep the inmates in good health and free from vermin, the cage should be thrown aside.

Never go out for a walk without bringing home a green sod for your skylark, and a little groundsel, chickweed, or plantain for the rest of your birds; who can tell but what a little fresh green-meat hung tastefully about their cages, causes them to fancy that they are once more in their natural haunts. Never let sand nor water be wanting.

ENGLISH TALKING BIRDS.

You have most of you heard at one time or another some of those wonderful stories which are told of talking birds, how a magpie, hanging in his cage near the docks, having listened to the carmen, learnt to call out, "Back, back, gee whoo-up," until one day he, by his calling, backed a cart into the dock, where both the horses, who had obeyed the voice of the mischievous waggoner, were drowned. Of the starling, whose cage hung opposite the little stall of a poor snob, that was wont to give a long whistle, then call out "Snob" every time the poor little cobbler went home with a job, or came back with old shoes to repair, and how the snob at last got so enraged that he used to come out of his stall, shake his fist at the starling, and call the bird everything but a gentleman, while the only reply was a more prolonged whistle, and a louder cry of "Snob" from the starling. How, one day, after a long altercation with the bird, the snob came out to shy his lapstone at it; how he missed the cage, and sent the lapstone through the parlour window bang into a large aquarium, which it smashed to atoms, and left such a lot of anemones, hermit-crabs, prawns, shrimps, and we know not what besides, sprawling over the carpet, as gave the parlour "a most ancient and fish-like smell" for weeks after; and as the value of the aquarium, with its contents, was some ten pounds or more, the snob, when he heard of the damage he had done, packed up his last and awl, never came back for his lapstone, and for aught we know, plunged into the first aquarium he came to, where he may be still swimming about even to this day. Then there was the raven, who, because the mistress was niggardly and kept the larder locked, the servants, unknown to the mistress, had taught it to call out, "Look to the cupboard; missus," and how one day the mistress came unexpectedly into the kitchen, and hearing the raven cry out, "Look into the cupboard, missus," that she opened a closet door and there found a great long-legged policeman, with half a loaf of bread in one hand, half a ham and a large knife in the other; and how, when ordered to come out, he was unable to speak on account of his mouth being so full of bread and ham. And that was where the servant girl had hidden her sweetheart when she heard her mistress coming into the kitchen, nor would the girl ever have been found out had she not taught the raven to say, "Look to the cupboard, missus." But these are only one or two out of the many scores of wonderful stories told of the things said and done by our English talking birds.

We have but five native birds that can be taught to talk, or rather to imitate the tone of the human voice, just near enough to tell what sounds they do utter, though it requires some little stretching of the fancy even to do this at times, and these are the raven, magpie, jackdaw, jay, and starling. What he did to these birds, or whether he did anything or nothing, beyond taking the twopence he charged for the (supposed) operation we never knew, nor ever shall know now; but in our boyish days an old man named Shaddy used to pretend to cut the tongues of talking birds, and we invariably took the young ones to him to be operated upon. He would never let us see what he did, for as he used to say, "That's my secret, and you would be as wise as I am were I to let you know it." In giving us back our birds after having got the money, he would sometimes say, "That jackdaw, after a little practice, will be able to preach in any church in England," or "That magpie will defend a prisoner as well as any counsellor that ever wore a wig." After cutting one of our raven's tongues, or pretending to do so, he gave us back the bird, saying, "Take care of that bird, there's a look of Shakspeare about him, and there's no knowing in time what he may do." I believe now that Shaddy was a regular old humbug, for the birds whose tongues he pretended to slit, never talked a bit better than the birds reared by other boys who kept their money and never went near him.

The MAGPIE takes the first place amongst English talking birds, and he will make himself heard, we can tell you, for his voice is so sharp and shrill that it almost seems to go through you. But then he is such a thief, and would steal and hide his old father if he didn't keep a sharp look-out. As for eating, we hardly know what he won't eat, excepting the coal-scuttle, and we believe he would have a try at that were it not too big; he eats bread and cheese like fun-o, and we have seen him try at a pot of porter, but that he didn't seem to relish much. Whatever you eat yourself, or nearly so, Maggy will eat, for nothing seems to come amiss to him: he requires a good-sized cage, and ought also to be allowed to run about as much as possible.

The RAVEN never ought to be put in a cage at all, and those who have not got an outhouse for Ralpho, and plenty of room for him to run about in, never ought to keep a raven. Like the magpie, it will eat anything it can get down its throat, even the mortar out of a wall when there is nothing better to be had. It can imitate almost any sound it hears, and is by many considered a better talker than the magpie; it is also equally mischievous, and when it has accomplished its purpose and is discovered, is apt to go off with a triumphant cry, as though it quite enjoyed the fun. It has a peculiar side-long step, and when alarmed, makes noise enough to startle every sleeper in the neighbourhood for a long mile round.

If the JACKDAW is not a good talker, he will make noise enough for a dozen birds, and thinks "no small beer of himself." Jack ought to have his wings cut, and be allowed to hop about, for he is a most amusing old-fashioned fellow. Then it is impossible to expect him to remain silent, living as he has done up in the old church tower among the clanging of bells that must have shaken every feather in his body while he listened to them. You should see him kill a mouse—just one tap of his thick, hard, horny beak, and Mr. Mouse is done for. A little bread and meat, or any odds and ends, seem to satisfy him. Some sounds he can imitate capitally, and one we knew so closely imitated the cry of a milkman as to bring the servant maids up the area steps with their jugs, when "Milk-o" was two or three street off. His hearing must have been very acute, for the milkman generally appeared some five minutes after the jackdaw commenced crying "Milk-o," and no doubt the bird heard him long before anybody else did.

The JAY is a beautiful bird, and quite an ornament to a garden. He can imitate a number of sounds, such as that of a man drawing a cork, the mewing of a cat, the bleating of a lamb, and the sounds of a few words, though never very distinctly. The blue markings on the jay's wings are richer than those of any other English bird. The jay will eat most kinds of grain, and when caged, nothing better can be given it than bread, a few acorns, and plenty of wheat.

The STARLING, in spite of all Sterne says to the contrary, seems to make itself quite as comfortable in a cage as most birds. The same food as that given to the woodlark seems to suit this bird as well as any when in confinement. It is also fond of meat. The bird is prettily marked, but is of no great value either as a singing or a talking bird, and soon forgets all it has been taught. Old Shaddy used to say that a starling might be taught to talk as many languages as it had slits made in its tongue, and that with a like practice it soon becomes perfect in Hebrew, Greek, and Arabic.

PARROTS

are not native talking birds, but as they do attempt "to murder the Queen's English" at times, with their head-aching screams of "Pretty Polly," and such like, which sounds as if they wanted a piece of fat bacon pulling up and down their throats with a string to clear their voices; and as they are such favourites with those good old aunts and maiden ladies, who "tip" so handsomely when they come to see us at school, we must say a few words about how they are to be fed and managed. Parrots require large cages, for when in health they are restless jades, and seem ever upon the move; the perches should also be thick as the grasp of their claws is rather large. Polly must also have a ring to swing upon; and when you have procured her all these comforts, and put your face to the cage for her to give you a kiss, perhaps the hooked-nosed traitress will nip off the end of your nose and disfigure you for life. Their food should never be placed in metal pans, but in either earthen or glass vessels. They are fond of bread or biscuit soaked in milk, especially if given to them before it is quite cold; boiled Indian corn is also excellent diet for parrots. They will eat almost any kind of wholesome seed or grain, and are very partial to nuts. The grey parrot, which is about the size of a pigeon, is considered the best talker, though the green parrot bears away the bell for beauty. By-the-way, we may as well tell you that the hackneyed phrase of "bearing away the bell" originated in the prize given in former times to the winner of a horse-race, which instead of consisting of a gold or silver cup, as is now given, was generally a silver bell.

Parrots require to be kept very clean: and although they will eat it readily enough, animal food ought never to be given them. The bottom of their cage ought always to be strewn with sandy gravel, which should be changed at least every other day. Holstein says, "The ugly brutes ought to be supplied with small-toothed combs, and compelled to use them," but he hated parrots: there is something on record about his wringing the neck of one because he could not study for the noise it made, and that his aunt left him the stuffed parrot as a legacy and nothing besides, so you see why he wrote so bitterly about parrots. They are certainly not the pleasantest companions when we are suffering from headache, or when wearied and ill, and having passed a sleepless night through the toothache, we are just sinking into a gentle doze, then we wish the cry of "Pretty Polly" was sounding over the Red Sea, and think the best food that can be given them is arsenic soaked in prussic acid. But there is one comfort after all, they are *not* English talking birds.

DOMESTIC FOWLS.

"A yard she had with pales enclosed about,
Some high, some low, and a dry ditch without.
Within this homestead lived, without a peer,
For crowing loud, the noble chanticleer;
High was his comb, and coral-red withal,
In dents embattled, like a castle wall:
His bill was raven-black, and shone like jet;
Blue were his legs, and orient were his feet,
White were his nails, like silver to behold,
His body glittering like to burnished gold."

THE beautiful lines above, descriptive of the old English farm-yard cock, were modernized by Dryden and not at all improved, if compared with the original, written by Chaucer, author of the " Canterbury Tales," who was buried in Westminster Abbey in the year 1400. It is beyond doubt the finest word-painting of Chanticleer ever penned by poet. Poultry are not only pretty, but profitable. Every boy is fond of a new-laid egg now and then, and what a smile there is on the mother's or sister's face after it has been decided that a pudding shall be made for dinner, when you bring in the basin of eggs unexpectedly, having more than you require the hens to sit on. Then it is an interesting sight to see the hen with her little brood of chicks, to notice what care she takes of them, and, above all, to know that Our Saviour looked upon the same objects and said to a rebellious race, " How often would I have gathered thy children together, even as a hen gathereth her chickens under her wings;" but He tells us they would not come. Then think of the beautiful passage about the hen lifting up her head to heaven every time she drinks, in Bunyan's " Pilgrim's Progress;" but were we to point out a thousandth portion of the descriptions of these beautiful domestic birds, we should fill the whole of the space allowed for our present brief article.

COMMON FOWLS

are supposed to be of eastern origin, although we doubt it very much, for if not brought over by the Romans when they invaded England, they were well known to the early Saxons soon after their departure, and also to the ancient Britons anterior to their time. Every boy knows what splendid-looking creatures the bantam cock and hen are, and has seen them so often, that we need not dwell upon the graceful fall of the hackles of the cock, sometimes looking like a golden cape thrown over his neck, the fine arching sweep of his tail, his tiara of comb, and the proud round sweep of his breast, for every boy has seen and admired him many a time and oft.

GAME FOWLS

are not set so much store by now as they were in the brutal days of cock-fighting, bull and badger-baiting, all of which cruel and disgraceful recreations, or sports, as they were wrongfully called, have passed away. The game-cock, especially the red, is a gorgeous bird, and beautifully formed, and so brave, if thorough-bred, that he will never run, but fight till he dies, and even in the death struggle try to lift up his armed heel or peck at his conqueror.

DORKING FOWLS

are well-known Surrey birds, valued most when of a pure white, and readily known through having five claws on each foot. They not only lay splendid eggs, but have quite a pleasant appearance when either boiled or roasted, the " sniff" of which is delightful to a hungry boy, who, forgetting all about their plumage, seizes his knife and fork and defaces their neat anatomy, without having even the courtesy to say "by your leave." They are among the largest of our common poultry.

THE BANTAM,

or DWARF COCK, is the smallest of all gallinaceous birds, but in pugnacity and pluck is equal to most, as it will fight to the last with one much larger and more powerful than itself. Its tiny size, pretty plumage, and high-spirited bearing, as it struts

"Royal as a prince is in his hall,"

have made it a very general favourite, and caused its elevation to the rank of a fancy bird. The rules respecting the colours and sorting of the feathers, general carriage, and other properties, as settled by fanciers, are the following:—For colours, nankeen and black are the most prized; if the bird is of the first colour, the edges of his feathers should be black, tail feathers black, breast feathers black with white edges, wings barred with purple, and his hackles or neck feathers slightly tinged with purple; and if of the second colour, no feathers of any tint should break the uniform yellow tone;

F

in carriage he should be free and spirited, and in general properties he should have a rose comb, full hackles, and full-feathered tail, and his legs quite clean, bright in colour, and wholly free from feathers; in weight he must not exceed a pound. The hens must be small, and correspond in plumage with the cock, and like him be clean-legged. Such is considered a genuine fancy bird. Those which are called after Sir John Seabright are another beautiful variety, and there are a third variety, which are spotted and streaked like a partridge, while the hens lay pheasant-coloured eggs.

POLISH FOWLS

are of a brilliant black colour, with white toppings on their heads, which are flat and surmounted with a fleshy appendage, from which the crown-feathers spring. They are more trouble than common fowls, more subject to disease, and require more warmth. Their top-knots require clipping, or they fall down if allowed to grow too long, and prevent the birds from seeing.

MALAY FOWLS,

generally called Chittagong, are an Indian variety; large in size, in colour yellow streaked and dark brown, have long necks, are small-headed, and stand high on the legs. The hens, if well fed, lay immense eggs.

SPANISH FOWLS.

Every boy knows these beautiful birds, with their black plumage, loose comb often hanging gracefully down. For gaudy colouring, the gold and silver-spangled bear the palm; but they are not pure Spanish, and have been mixed with the Polish fowls, though when fine they are considered of great value. Spanish hens, though they lay large eggs, seldom sit well.

PERSIAN FOWLS.

These lay well, but are nothing to look at, having no tails, and looking like poultry in boy's jackets, and never to be honoured with a lappel coat. They are ugly enough to be eaten.

COCHIN-CHINA FOWLS.

These are monster birds, the cock often reaching the height of two feet, and the hen only some four or five inches under that standard. The buff and cinnamon-coloured are the greatest favourites, though the white ones at times realize immense prices. There was quite a rage for keeping these birds some few years ago, which of late has much abated, as every "nobody" took to rearing Cochin-Chinas, and many a curse did they call forth in the night in drowsy-headed neighbourhoods, as no one could get any sleep for their dreadful crowing; as they do not crow in English, but in the Chinese tongue, and their crowing sounds like something between the cross of a shriek and a whistle, such as Commissioner Yeh gave when seized by his pigtail.

HOW TO CHOOSE STOCK.

Always breed from young stock: pullets in their second year, and a stag or cock two years old, are the best ages to commence with; for a hen is in her prime when three years of age, and begins to decline after the fifth year. Never select a hen for sitting that imitates the cock in her crowing, and has a large comb; such a hen is of no use as a breeder. Yellow-legged fowls are seldom robust; those that have eyes which sparkle like diamonds, and combs red as a ruby, and are bold, fierce, and active, are the fowls to breed from, for these are sure signs of good health. Spring is the best time to lay in your stock, and one cock to nine hens is the best. Old fowls are stiff in the feathers—stumpy, as it is called—have large scales on their legs, comb and gills full, and of a dull dead whitey-red colour. Be sure, when purchasing stock, to look out for these marks, and have nothing to do with such birds.

HOW TO FEED FOWLS.

Fowls will eat either vegetable or animal food, when allowed to run about; they feed greedily on worms and insects of almost any kind, and would no doubt gobble up an alligator or a boa-constrictor, if they could get them into their crops, and "grind their bones to make their bread" in their gritty gizzards. Barley, oats, tares, peas, millet, and sunflower seeds, form their favourite food; they must also be supplied occasionally with green meat—grass, if nothing better can be had. They must always have plenty of clean water to run to whenever they please. They should be fed regularly twice a day. A handful of boiled potatoes and carrots mashed small, will at times, as the old countrywomen say, "do them a world of good."

Fowls should be permitted to range in the open air during the day as much as possible, and their habitations for the night be warm, dry, clean, and well ventilated; and there must be perches for the birds to roost upon, and also boxes, having fine soft hay or short straw inside, in which they may build their nests. A piece of chalk may be put into each box as a nest egg, and it is necessary to take the real eggs away as soon as they are laid. When some of the hens by clucking evince a desire to sit, they should be kept in a box apart from their companions, with from five to nine, or at most eleven, eggs to hatch. Old nests should never be used, and the boxes in which the birds are put up must be clean. Incubation continues for twenty-one days, and during that period food and water must be placed near the nests, that the hens may eat and drink without having to go far for the purpose, so that the eggs may not chill. The food proper for the little chicks consists of split grits, chopped curds, and eggs boiled hard and cut into very small pieces; as they increase in size, they should gradually be brought on to eat the same food as full-grown fowls, which is tail-wheat, barley, oats, &c. Water must be furnished them in little shallow pans, so that the chicks may drink without hopping into the water, and so wet their feet and feathers; for when young, such a cold bath is apt to numb and injure them.

The diseases to which chickens are liable are the chip, pip, and roup. When suffering from the chip, the little things sit moping and chipping in corners, and seem to be perishing with cold; warmth, and some mustard or pepper put into the water are the best, and in fact the only, remedies. The pip is a white skin growing upon the tip of the tongue; it may be cured by scratching the skin off with the nail, and rubbing the place with salt. The roup is another disorder which requires warmth to counteract its effects; the bird's nostrils should be washed out with warm water, and some pills, composed of butter and chopped rue leaves, administered every day. Full-grown fowls are sometimes attacked by this disease, and it not unfrequently proves fatal.

PIGEONS.

WILLIAM BROWN, who wrote those beautiful poems in the time of Queen Elizabeth, entitled "Britannia's Pastorals," speaking of the colour of a dove's neck, says no one can tell where the blue and purple begins or the green ends. Nor is there a more graceful object in creation than a beautiful dove. The horse is not more elegantly formed in that fine sweep from head to back which makes Hogarth's "line of beauty," or part of the letter S. Then no queen can move more stately than the proud pigeon with his head thrown back and his breast thrown forward, walking as if he were the lord of all creation, and fully conscious of his own beauty. What an eye he has, too; there never was a precious stone discovered in the world to equal it: it has the liquid light of the diamond, the fiery blaze of the ruby, and as for plumage, all the colours of the rainbow and all the shades of all the flowers that ever blowed, may be found in the dove's neck alone. What a lover of doves King Solomon must have been, and how beautifully he alludes to them, and how his heart rejoiced when he spoke of the voice of the turtle being heard again in the land, and said of his lovely queen, "Thou art fair, my love, thou hast dove's eyes," calling her his "dove in the clefts of the rocks;" then he speaks of "doves by the rivers of waters," and of his "dove being the only one of her mother;" and it is pleasant to know that there were pigeon-fanciers and a cooing of doves in the ancient streets of Jerusalem—those streets which on a later day our blessed Saviour trod. From time immemorial they have been the emblems of impassioned love and faithful attachment, and the fidelity of the turtle-dove to its mate has been sung by our great Shakespeare and almost every other poet, and is now proverbial.

THE STOCK DOVE, OR WILD PIGEON.

The stock dove, or wild pigeon, is supposed to be the original stock from which the different varieties of the domestic pigeon are derived, but this has never been clearly proved. It is about fourteen inches in length, and in plumage is exceedingly beautiful. It is one of the three species that live wild in our country, the other two are the turtle-dove and ringdove, the latter being the largest of the three, and so called from the black ring round its neck, which is edged with white. The head, neck, and upper part of the back of the stock dove are of a deep bluish-grey and purple colour, reflected on the sides with green and gold, and that so delicately, as caused William Browne, the poet, to write,

"That none can say, though he it strict attends,
Here one begins, and there another ends."

Its breast is a faint reddish purple, belly, thighs, under tail coverts, and the lower part of its back and rump, a light grey, or ash colour; primary quill feathers dusky, edged with white; the others grey, marked with two black spots on the exterior webs, so as to form two black bars across each wing; its bill and legs are red, and its claws black. Stock doves are migratory birds, visiting England in large flocks at the beginning of November, and retiring at the end of the spring, though some remain with us, like the ringdove, all the year, and only change their quarters in search of food. The stock dove builds a slovenly nest of sticks, which can be seen through, and lays two eggs.

THE TURTLE-DOVE

is one of the prettiest of the species; its bill is a bluish-brown, eyes yellow, and surrounded by a crimson circle; the back of its head is of an ashen-gray colour, and on each side of its neck is a patch of black feathers, margined with white; its back is ash colour, and each feather tipped with reddish brown; wing coverts reddish brown, spotted with black; quill feathers dusky, with light edges; the throat, neck, and breast tinged with a beautiful red; the two middle feathers of the tail brown, and the others dusky, tipped with white, and its legs red. They visit England in May, and leave in September. Young birds reared by domestic pigeons soon become accustomed to the dove-cote, but as they are very susceptible of cold, they require to be protected from the chills of winter. Turtle-doves may be fed with any kind of grain, peas, beans, and such like. They soon become tame in confinement, and caress those who feed them. The best thing to keep them in is a warm cage; there are cages made purposely for them.

THE COMMON PIGEON.

Common pigeons are generally blue or ash coloured, with white backs and red legs; but by paying attention to the crossing of breeds, their plumage may be enriched with tinges of copper, yellow, and other lively colours. They require very little care, and are very prolific, breeding seven or eight times a year, laying two eggs each time, which generally produce a male and female; and it is amusing to watch the eagerness of the male to sit on the eggs while the female rests and feeds herself. The young, when hatched, require no food for some time but what they receive from the female.

THE FAN-TAIL, OR BROAD-TAILED SHAKER.

This beautiful variety of the pigeon tribe receives the name of FAN-TAIL from its habit of spreading out the feathers of its tail like a turkey-cock, and that of BROAD-TAILED SHAKER from its breadth of tail, and a peculiar quivering motion of its neck. It has a full breast, and a short, handsomely formed, arched neck, which it carries in a graceful, swan-like curve. Its tail, according to the rules of the fancy, should consist, at the least, of twenty-four feathers, and at the most of thirty-six, which number it should not exceed, for if the tail is over-crowded with feathers, the bird suffers it to droop, a defect never passed over, although the bird may be faultless in every other respect. Fantails whose plumage is pure white are more highly prized than those displaying red, yellow, blue, and black-pied colours, their carriage of the neck and tail being considered by far the most striking and elegant.

THE NARROW-TAILED SHAKER.

Some fanciers are of opinion that this bird is a breed between the broad-tailed shaker and the stock-dove, whilst others imagine that it is a distinct species. Its neck is shorter and thicker, back longer, and it has not so many tail-feathers as the broad-tailed shaker, neither does it expand its tail so fully, keeping the feathers rather closed one over the other, so as to resemble a fan when some little way open. The colour of its plumage is generally white, but a few different tints, and even an almond variety, are to be met with occasionally.

THE DUTCH CROPPER.

This species of pigeon is gravel-eyed, and thick, short, and clumsy in the body and legs, which should be feathered down to the feet. These birds have a large crop or bag under their beak, which they can inflate with wind, or depress at pleasure, and they are such careless parents, taking so little heed of their young ones, that it is requisite to put the little things, as soon as they have fed off their soft meat, under a pair of dragoons, pouters, or small runts. Care must be taken to supply the croppers regularly with food, else they will gorge themselves—a habit they are extremely addicted to unless properly tended.

THE ENGLISH POUTER, OR POUTING HORSEMAN.

This fancy pigeon was originally bred in England, and thence derives its first name, and from being a cross breed between a horseman and a cropper, its second title; through judicious pairing with the cropper, it has attained great beauty and high value. Pouters are very expensive birds to rear, as the strain will soon become degenerate and worth nothing; the fancier will, therefore, even if he commences with a stock of several pairs, be often obliged to sell or exchange really good birds for inferior sorts, in order that he may be enabled to cross the breed. As the old birds pay little attention to the wants of their young ones, it frequently happens that they are starved to death; careful fanciers, therefore, never allow them to hatch their own eggs, but "ring the changes," by putting them under a hen dragoon that has lately laid, and placing the eggs of the latter bird under the pouter, in order that the pouter may commence incubation, otherwise she will lay again in a short time, which, often repeated, would in all likelihood kill her. Every bird must be kept by itself during the winter season, and their coops must be lofty, so that they may not acquire a stooping habit, which is a very great fault. In the spring every pair of pouters must have two pairs of dragoons to tend and feed them, but care must be taken that the dragoons are kept in a loft separate from the pouters, else a cross breed may probably be produced, and the stock become degenerate.

The rules laid down by the fanciers regulating the various properties which a first-rate pouter should possess, are—from the point of the beak to the tip of the tail the bird should measure eighteen inches; its shape should be fine, and its back hollow and tapering from the shoulders, for if there is a rise in its back, it is termed hogbacked, and therefore considered defective; it should carry the shoulders of its wings close to its body, and display the wings without straddling. Its legs, from the toe nail to the upper joint of the thigh, should be seven inches in length, stout, straight, and well covered with white soft downy feathers, not marked with any other colour about the thighs or knee-joints, which is termed foul thighed. The crop ought to be large and circular towards the beak, and rise up behind the neck, so as to cover and run gradually off at each shoulder; the bird should fill his crop with wind, to show its full extent, with ease and boldness; but if the bird does not fully inflate his crop—that is, only just enough to make himself look like a badly made runt, he is termed loose winded. In point of colour, the blue, black, red, and yellow pieds, are the most esteemed; but if a blue pied and a black pied are equally fine in their properties, the black pied, on account of its plumage, is the most valuable of the two; and if a yellow pied has the same markings as the two former, it will be more prized than either. The manner in which the markings ought to be distributed over the bird is the following: the head, neck, back, and tail should be uniform in tint; a blue-pied pigeon must have two black streaks near the end of both wings, but if the stripes incline to a brown tint the bird is termed kite barred, and its value is thereby greatly deteriorated; when the pinion of the wing is speckled with white, in the form of a rose, it is called a rose pinion, and is much prized; and if the pinion is marked with a dash of white on the outer edge of the wings, the bird is reckoned bishopped, or lawn sleeved. If the nine flight feathers of the wings are not white, the bird is foul flighted, and if the outer wing feather only is white, it is sword flighted. The front of the crop should be white, surrounded by a shining green, interspersed with the same colour with which the bird is pied; but the white must not reach so far as to pass round the back of the head, for then it would be considered a ring-headed bird; upon the crop there should be a crescent-shaped patch of the colour with which it is pied, and when that is missing it is termed swallow-throated. Pouting horsemen are not so much in repute as formerly, the almond tumblers having almost superseded them in the estimation of the fancy.

THE PARISIAN POUTER.

This species was introduced, as its name implies, from Paris; it is short in its body and legs, thick in its girth, and has generally a long but not a very full crop. Its plumage is much admired, as every feather—the flight feathers excepted, which are white—is elegantly streaked with a rich variety of colours; if much red is intermixed with the other colours, the more valuable is the bird considered. They are usually what is termed bull or gravel eyed.

THE UPLOPER.

This bird was originally brought from Holland; in appearance it greatly resembles an English pouter, only that it is somewhat smaller in all respects; it has thin legs, and its toes are very short and close together, and it tips so exactly upon them when walking, as to leave the ball of the foot quite hollow; its crop is very round, and it generally hides its bill amongst the feathers upon it.

THE HORSEMAN.

Many fanciers suppose that the horseman is a cross-breed, either between a tumbler and a carrier, or a pouter and a carrier, and then again bred from a carrier. In shape it resembles the carrier, but it is smaller in all its properties, its body being less, and its neck shorter, and the fungous-looking flesh round its eyes not so exuberant, so that there is a greater space between the wattle on the beak, and that round the eye. The most approved colours for this species of pigeon are the blue and blue pied, as they are usually the best breeders. They should be flown twice a day regularly when young, and as they gain strength on the wing, be allowed to range loose, without any other birds in company. This variety of the pigeon tribe is the kind most generally employed in carrying letters, the genuine carriers being much too scarce and valuable to be commonly used.

THE TUMBLER.

These pretty pigeons derive their name from a peculiar habit of tumbling backwards in the air when on the wing; besides which, in addition to their tumbling acquirements, they soar to so great a height as to be almost lost to the view; when flying they congregate very close together, and if good birds and accustomed to each other, they will maintain such a compact flight, that a dozen may almost be covered with a large handkerchief. If the weather is warm and bright, they may be allowed to wheel about and gambol on the wing for four or five hours in succession; but care must be taken that no other species of pigeon mixes with them, for if they once become familiarized and fly with others, they will gradually drop their highly-prized mode of flight, and become worthless. They should never be let out on a dull, heavy, misty morning, nor when a fog appears to be rising, nor during a high wind, as all such atmospheric variations, by causing desertions from the *aërie*, tend to diminish the stock. A hen tumbler should never be suffered to fly while with egg. The most esteemed tumblers do not summersault when swooping along, but only when they are beginning to rise, or when coming down to pitch, and to preserve this and the high-flying properties in his stock, the provident fancier must spare no expense in the purchase of one or two first-rate birds that have been used to soaring, as they will be of much service in training the young ones. When the birds are accustomed to their habitations they may be turned out, and put upon the wing—once a day only, however; a bright grey morning is the best time, especially for young birds, and some hemp or canary seeds must be scattered round their cotes to entice them in.

There is a particularly fine variety of this species, called the bald-pated tumbler, from its having a beautiful snow-white head; it has pearl eyes, and in plumage is exceedingly diversified, the tail and flight feathers, however, matching the head, being of a pure white. When a tumbler, either of a black or blue colour, has a long dash of white from the under jaw and cheek to a little way down the throat, it is called a black or blue-bearded bird, as the colour may be; and if this beard is well shaped, and the bird is clean in the tail and flight, as before described, it may be reckoned very handsome and valuable. When these birds are dashing along in the brilliant sunshine, the lively contrast of their feathers adds much to the beauty of their appearance, as every hue is then touched with golden splendour.

There is another and still more beautiful variety of the tumbler, styled by some fanciers the ermine, but which is most generally known by the name of the almond tumbler. It is an extremely elegant and highly-prized variety, and derives its origin from common tumblers judiciously matched, such as yellows, duns, whites, black splashed, black frizzled, &c., so as to sort the feathers. When in perfection, tumblers are esteemed by many persons as the prettiest of all the pigeon tribe, and this high opinion is borne out by the beautiful diversity of colours which they show, which is so elegant and rich in some birds, that they have been compared to a bed of tulips. The more they are variegated in the flight and tail, especially if the ground colour is yellow, the more they are valued; and the reason why a fine bright yellow ground has the precedence of all others, is from its being so exceedingly difficult to acquire, as twenty light-coloured birds may be procured for one displaying a deep, richly-tinted ground. There are also the black-mottled and yellow-mottled tumblers, named principally from their colours only.

THE RUNT.

There are several species of runts—the chief of which are known as the common, or dove-cote, the Roman, the Smyrna, the Friesland, the Leghorn, and the Spanish. The common species are usually good nurses, and are generally employed in that capacity for the more valuable kinds of pigeons. The Roman runts are so large and unwieldy, that they are scarcely able to fly; those of Smyrna are middling-sized birds, with feathers sprouting from the outside of their feet, so as to present the appearance of small wings. The runt of Friesland is rather larger than the middling-sized common runt, and its appearance is very singular from its feathers being all inverted, or turned the wrong way. The Leghorn species is a fine full-bodied, short-backed, broad-chested, close-feathered pigeon; its head

is shaped like that of a goose, it is hollow-eyed, and round the eye is
a circle of thin skin; its beak is very short, with a small wattle over
the nostrils, and the upper chap projects a little beyond the under;
when walking it raises its tail up like a duck. The birds of this
species are much more hardy than some fanciers imagine, and breed
pretty well; but as they make very indifferent nurses, they should not
be trusted to bring up their own young ones; their eggs must therefore
be shifted under a dragoon, or some other tender nurse, taking care
to give them a young bird of some other variety to attend to, in
order to take off their soft food. Their plumage is usually of a
grizzled colour, ermined round the neck; but the birds most prized—
and none of them are beauties—are those which are either red, white,
or black mottled. The Leghorn variety is of greater value than any
other kind of runt, and by judicious crossing with the Spanish breed
a breed may be obtained of large size. Spanish runts are short,
thick-legged, flabby-fleshed, loose-feathered birds, having very long
bodies; their plumage cannot be criticised by rule, as it is particularly various; the best, however, are said to be those which are
either of a blood red, or mottled tint.

THE FRILL-BACK

is remarkable only for the peculiar curl of its feathers, which are so
turned at the end as to make a little hollow in each of them; it resembles the runt in shape, but is smaller than that bird; its plumage
is pure white.

THE CARRIER.

The carrier is somewhat larger than
most of the common pigeons; its feathers
lie very close and smooth, and its neck
is long and straight. From the lower
part of the head to the middle of the
upper chap, there is a lump of white,
naked, fungous-looking flesh, which is
denominated the wattle; this in good
birds is met by two small swellings of
similar flesh, which rise on each side of
the under chap, and if this flesh is of
blackish colour, the bird is considered
very valuable. The circle round the
black pupil of the eye is usually of a brick-dust red colour; but if it
is of a brilliant red tint, it adds considerably to the value of the bird;
this circle is surrounded by another of naked fungous flesh, generally about the breadth of a shilling, but when of greater breadth
the more it is admired. When the incrusted flesh round the eye is
very thick and broad, it shows that the pigeon will prove a good
breeder, and one that will rear fine young ones. The properties
attributed to the carrier, and prized by the fanciers are, three in the
head, three in the eye, three in the wattle, and three in the beak.
The properties of the head consist in its being flat, long, and straight;

as for instance, if the head is very long, narrow, and flat, it is reckoned, as far as shape is concerned, perfect; if the contrary, it is termed a barrel head. The properties of the wattle of the eye are its breadth and circular and uniform shape, for if one part appears to be more scanty than another, it is termed pinch-eyed, and is of comparatively little value; while if it is full, even, and free from irregularities, it forms a rose-eye, which is highly prized. The wattle should be wide across the beak, and short from the head to the point of the beak, and lean a little forwards from the head, as the bird is said to be peg wattled if it lies flat. The beak must be black, long, straight, and thick; if it is an inch and a half in length, it is considered a long beak, but it must never measure less than an inch and a quarter; if the beak is crooked, or, as it is termed, hook-beaked, or a thin spindle beak, the value of the bird is much diminished. This species is in general either dun or black in colour, although white, blue, splashed and pied specimens occur; the black and dun birds are usually the most perfect in their properties; but as the blues, whites, and pieds are very rare, inferior birds of these colours are of considerable value. The carrier has been termed king of the pigeons, from the elegance of its shape, and great sagacity. In wager-matches these birds have flown from London to Antwerp, or to Paris, in less than six hours. The telegraph will soon render them useless; but in early times they were the messengers sent in cases of life and death.

THE MAWMET.

The mahomet, commonly corrupted to mawmet, and supposed to belong to that species of pigeon which aided the great impostor after whom it is named, by being trained to approach his ear, is a beautiful cream-coloured bird, with bars of black across its wings; and although the outside or surface of its feathers is of a cream-colour, yet the part next the body, the flue feathers, and even the skin, are of a dark sooty tint; it is much about the size of a turbit, but instead of a purle it has a fine gullet, with a handsome seam of feathers; its head is thick and short, eyes orange-coloured surrounded by a small naked circle of black flesh; it has also a little black wattle on its beak.

THE BARB.

This species was originally introduced from Barbary; in size it is a little larger than the jacobine: it has a short thick beak, a small wattle, and a circle of thick naked incrusted flesh round its eyes; the wider this circle of flesh spreads round the eye, and the more brilliant it is in colour, the more highly the bird is prized; this circle is narrow at first, and is not fully developed until the bird is three or four years old. The plumage

of the barb is usually either dun or black; but there are pied birds of both colours, which are held in but little estimation, as they are supposed to be only half bred. Some of this species are ornamented with a tuft of feathers rising from the back part of the crown of the head.

THE SPOT,

Is so called from a spot of colour just above the beak. Its body is mostly white; the tail feathers generally correspond in colour with the spot, which is either red, yellow, or black, and sometimes, but not very frequently, blue; they invariably breed young ones of their own colour.

THE DRAGOON.

Dragoons are a breed between a tumbler and a horseman, and by frequently crossing them with the horsemen they acquire much strength and swiftness. They are exceedingly good breeders and make kind nurses, and are therefore often kept as feeders for rearing young Leghorn runts, pouters, &c. The dragoon is rather lighter and smaller than the horseman, and one of its chief beauties consists in the straightness of the top of its skull with that of its beak, which, according to the rules of the fancy, should almost form a horizontal line. These birds should be flown and trained while young, in the same way as the horsemen, which they are considered to surpass in swiftness in short flights of from ten to twenty miles; but in longer distances, if the horsemen are well bred, they will far outstrip the dragoons.

THE JACOBINE.

This bird, when perfect in its properties, is scarce. The real jacobine, or, as it is most frequently termed, the Jack, is a very small pigeon—and the smaller it is, the more valuable—with a range of inverted feathers on the back of its head bending towards the neck, somewhat like the hood or cowl of a monk, and from this peculiarity it receives the name of jacobine or capper. These feathers are termed the "hood," and if they are compact and grow close to the head, they enhance the value of the bird considerably; the lower part of the hood is called the "chain," and the feathers composing it should be long and thick.

A small head, very small spindle-shaped beak, and beautifully clean pearl eyes, are other properties of this little pet. Yellow, red, blue, and black, are the colours most usually bred, and in point of colour the yellow birds are preferred before all others. According to the rules of the fancy, the tail, flight, and head must invariably be white. Some of these birds have their legs and feet covered with feathers.

THE CAPUCHIN,

In its properties, is closely allied to the jacobine, indeed, some fanciers consider it a mixed breed between a jacobine and some other kinds; whilst others affirm that it is a distinct species of pigeon. It has a longer beak, and is altogether larger than the jacobine; it is destitute of the chain, but its hood is extremely pretty.

THE RUFF.

This bird has sometimes been passed off as a jacobine, but it is a distinct variety, as it has a bigger head, longer beak, and is a larger bird. The chain of feathers does not flow so near the shoulders of its wings, and neither that nor the hood are so compact and close as those of the jacobine, although the feathers of which they are composed are longer.

THE LAUGHER.

This bird is of Oriental origin, and was first brought into Europe from Palestine. In size and shape it resembles a middling-sized runt; the colour of its plumage is usually a mottled red, and sometimes blue. Its eyes are very brilliant, clear, and pearly in tint, approaching to a white. It derives its name from a singular note which the cock utters when he seeks his mate, which greatly resembles a laugh.

THE TRUMPETER.

This variety of the pigeon tribe is almost as large as the runt, and resembles it in shape. The crown of its head is round, and the larger it is, the more it is prized; a tuft of feathers rises from the root of the beak, and according to its size so is the value of the bird, and its legs and feet are covered with feathers. These birds derive their name from making a noise like the sound of a trumpet, and this they always do in the spring time of the year; those persons who wish to hear their trumpet-like call at other times, supply them plentifully with hemp-seed, which invariably has the desired effect. They are generally pearl-eyed, and in plumage black mottled. The trumpeter is not an especial favourite amongst true fanciers, notwithstanding its curious note, as it is usually classed amongst those pigeons which are called "Toys."

THE TURBIT.

This pigeon is rather larger than a jacobine; its head is round, beak short, it has a tuft of feathers growing from the breast, and spreading like the frill of a shirt; this tuft is termed the purle; and according to the size of this purle and the shortness of its beak, the bird is valued; it has also a gullet which reaches from its beak to the purle. Yellow, dun, red, blue, and black, are the colours most frequently seen in this variety of the pigeon tribe, but some few chequered ones are occasionally bred. According to the fancy, the back of the wings and tail should be all of one colour, the yellow and red-coloured birds excepted, whose tails ought to be white; a bar or stripe of black should cross the wings of the blue-coloured ones, but the other parts of the body and flight feathers must be white; they are termed yellow-shouldered, black-shouldered, &c. according to their different tints. When of a whole colour, such as blue, white, or black, these birds have often been sold as specimens of another variety, the owl. If well trained, when young, they will become very excellent flyers.

THE OWL.

The owl is somewhat less than a jacobine; it is gravel-eyed, and has a short curved beak, very much like that of an owl, from which it derives its name. The purle of the owl is rather larger, and in form much more like an expanded rose than that of the turbit; but in other respects, with the exception of the beak, the birds resemble each other so closely, that any further description is unnecessary. Great care should be taken that the breeding-places of these birds are secluded from observation, and dark, for the slightest alarm frightens them, and when molested, they fly from their eggs.

THE NUN.

The nun is greatly admired, from the elegant contrasts of colour in its plumage. Its body is generally white, and its tail and six flight feathers of its wings should be either wholly red, vivid yellow, or black, as likewise its head, which is adorned and nearly covered by a tuft, a "veil" of pure white feathers; according to its colours, the bird is termed a red, yellow, or black-headed nun, as it may happen to be; and whenever the feathers vary from this rule, the bird is of little value. The nun should have a small head and beak, a pearl eye, and if the tuft or veil is very large, the more beautiful and valuable the bird is considered.

THE HELMET.

The helmet is larger than the nun; its head, tail, and flight generally correspond in colour, being either of a yellow, blue, or black tint, and the other parts of the body are usually white; its head is ornamented with a delicate soft tuft of feathers, differing in colour from those of the body, and bearing a slight resemblance to a helmet, whence its name. Helmets are very pretty birds, but of no great value.

THE MAGPIE

Is a clean, handsome bird, resembling the magpie in colour, though it is often red, yellow, or blue, but black and white are the most valuable.

THE LACE

Is rather a scarce bird, about the size of a common runt, and much resembles it in shape; the colour is always white, and the webs or fibres are apparently quite unconnected with each other, seeming to be disunited throughout, which peculiarity makes the bird look singular yet pretty.

THE FINIKIN

Differs very little in shape from the runt, is snake-headed, gravel-eyed, and has a tuft of feathers growing on the back part of its head, which falling down the neck, carries, in some measure, the appearance of a horse's mane; it has a clean leg and foot, and its plumage is always blue or black pied. When cooing, its gestures are very curious, as it rises over its hen, flaps its wings, turns round three or four times, first one way and then another.

THE TURNER

Greatly resembles the finikin, but has no tuft on the back part of the head. It differs in its evolutions, as it turns round only one way.

THE DOVE-COTE.

To possess a really good flight of pigeons, a fancier ought to have out-buildings and other conveniences which might be made into nice roomy dove-cotes. Bird-fanciers, and persons who breed fancy pigeons, not having these conveniences, generally convert the lofts between the ceilings of the garrets and the tiled roofs of their houses into pigeon-lofts, and make an opening in the tiling, so that they are enabled to have a kind of platform outside on which they fix their traps or aëries; but such situations are often rather dangerous, for if an accident occurred it might prove fatal. The pigeon-fancier who is limited to a small space, must therefore be content with two or three, or at any rate a few, choice birds, and provide as capacious a cote as he can. The most common shape is the one represented in the annexed illustration, but the form of it is quite immaterial, the taste of the individual fitting it up, or the materials at hand, being better guides on those points than any rules we can place before him. It is necessary, however, that the holes should also be large enough for the birds to turn round in with ease, and shelves and partitions of from seven to nine inches in depth should be made along the front to keep the couples apart and afford them good resting-places; if two holes for each couple can be allowed between each partition it will be advantageous. The cote should be protected from the inroads of cats and rats, and this we have seen done effectively when the cote has only been a cask mounted on a pole, through a hole having been made in the bottom of an old saucepan, which was placed a yard or so under the dove-cote.

If the young fancier is allowed to fit up a loft over a stable or some similar out-building for a pigeon-house, the best arrangement he can adopt is that shown in the accompanying illustration. The means for exit and re-entrance must be first thought of, and if no window is in the loft, two holes must be made in the wall, at about five feet from the floor, each sufficiently large enough to admit a pigeon through easily. A shelf must also be fastened both inside and outside of these openings, and on the outside shelf a trap or aërie should be affixed, the intent and purpose of which we shall presently explain. At the upper part of the loft perches, made of rough branches, must be fastened for the birds to perch on, as shown in the engraving. At about seven or eight feet from the floor breeding boxes, according to the number of birds intended to be kept, should be securely fixed to the wall; and it must be borne in mind that it is necessary to put them at that height from the floor to protect the birds from rats, &c. Old egg-chests may be turned into very good breeding-boxes, but they must be partitioned off inside, so as to form separate places for the birds to nestle in. Some fanciers furnish their boxes with little earthenware pans, or small baskets, made for the purpose, for the birds to deposit their eggs in; some prefer the pans, others the baskets, as in the baskets the eggs are not so likely to be broken; but the pans, if supplied with straw, are decidedly cleaner than the baskets; the pans should vary in dimensions, according to the species of pigeon for which they are purposed. It is as well to put two pans in each little room, as the hens frequently go to nest again when their brood are about three weeks old, leaving them to the care of their mates. Instead of egg-boxes, shelves partitioned off, and having sliding fronts for the convenience of cleaning, are ofttimes used; and if the young fancier intends to keep pouters, the shelves should be fourteen inches in breadth, and at least twenty inches apart, that the birds may not acquire an ungainly habit of stooping, which depreciates their value considerably. The fountain, a large-bellied glass bottle with a tolerably long neck, for water, is an important item in the pigeon-house; it should be laid down on a small three-legged stool, and so placed that its mouth may descend into an earthenware pan, by which arrangement the water will trickle slowly into the pan, and cease the instant it reaches the level of the mouth of the bottle, and a continued supply of fresh water is thereby kept up; in lieu of a stool, two or three bricks will give the bottle the necessary elevation, and answer equally as well.

It is indispensable to the proper thriving of the birds that the loft and shelves be kept clean, and gravel strown on the floor; indeed, the gravel must on no account be omitted, as pigeons are exceedingly fond of picking it, and cannot without the aid of gritty matter digest their food thoroughly.

The aërie before mentioned, which is fastened on the shelf outside the loft, is a trap made of laths; it has two sides and a front only, the wall of the loft forming the back; the front and sides act upon hinges so that they may be thrown open and laid flat on the platform, as in the annexed figure A B C, 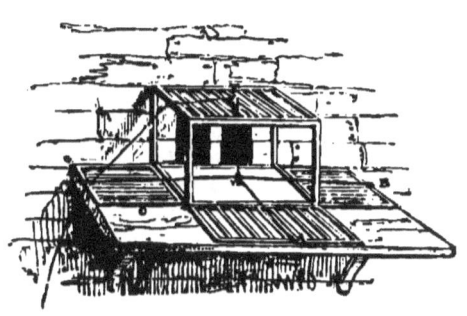 and on the upper parts of these flaps strings are fastened, which are united to a single string in the middle of the trap, the string is carried over the swivel E, at the top of the machine, and thence to a hiding-place, from whence the owner can see all that passes, and when a bird alights within the aërie he jerks the string, the flaps are elevated, and the bird is immediately a prisoner. The aërie, when shut, presents the appearance shown in the illustration. This kind of trap is used not only by fanciers, but by amateurs also, and is of great importance, both as a means of self-defence to secure strays and to shut in their own birds; among amateur fanciers the first-mentioned purpose is to

secure valuable and favourite breeds from being deteriorated through stray birds of no value pairing with them.

A bolting wire is recommended by some persons as a very useful addition both to the loft and the aërie; it is simple in its construction, yet efficacious, its chief use being to enable birds to get into the loft after the folding-doors of the aërie are shut; and indeed, if it is adapted to one of the doors of the aërie, it will prove eminently useful. An aperture sufficiently large to admit a pigeon must be made, and a slip of wood, nearly as long as the width of the opening, hung to the upper part of it by two small wire hinges, as shown in the accompanying illustration; this slip must move very freely, and into it two pieces of wire should be driven, the length of which must be regulated by the depth of the opening, taking care that they reach a little below the lower edge of it. The bolting-wire should of course only open inwardly, as the object of it is to let the birds in which may chance to be out late, and to keep them there when they have once entered, as they cannot open it from the inside, try all they may.

The pigeon call, by which the birds are enticed into their trap, or cote, or house, after they have been indulged with an hour's flight, is a very shrill, loud, and prolonged whistle; and if some favourite food is given to them, after they have attended to the call, they will by degrees become so well trained to it, as to respond to the signal whenever it is made. They should invariably be trained to come at this call before they are fed, and many persons, ere doling out their daily allowance of food, even when they are all in the loft, summon them together by it.

FEEDING.

Pigeons are fond of almost every kind of grain, but old tares are found by experience to be the best for them; horse beans, particularly the smaller sorts, such as small ticks, are considered next to tares in point of nutritive properties; oats, barley, buck-wheat, and peas may be given occasionally, and will be found salutary varieties of diet; buck-wheat, however, must be sparingly given. Rape, hemp, and canary seeds, pigeons are exceedingly fond of, but they must not be given with a liberal hand, and the same must be said with respect to new tares, which should, especially to young birds, be given very sparingly.

The seed may be scattered on the floor amongst the gravel, although many persons recommend the adoption of little contrivances to put it in, on the score of keeping it cleaner and better. Pigeons must be kept clean.

Pigeons make sad havoc of roofs and walls, by picking out the mortar, being very fond of lime and salt. To prevent this, take half a peck each of sifted gravel, brickmakers'-earth, and old wall rubbish, a pound and a half of cummin-seed, and a quarter of a pound of saltpetre, to be mixed with as much brine as will make it stiff; place this in anything where the birds can get at it easily.

MATING.

As mating or coupling pigeons is occasionally attended with difficulty, it is a good plan to build two cotes close together, having a lath partition between them, so arranged that the birds may see each other and feed out of the same little vessels, and by giving them plenty of hemp-seed they will soon be fit for mating. When you observe that the hen sweeps her tail, you may put her in the cock's pen, and they will readily agree. Where it is not convenient to make this probationary pen, and you are obliged to place them both in one coop, put the cock in a few days before his mate, that he may get accustomed to and feel himself master of it, especially if the hen is high-spirited, else they will quarrel so fiercely that their disputes will terminate in a total dislike of each other. When the pigeons are comfortably matched, they may either be allowed the full run of the loft, to select a nest for themselves, or fixed to one in particular, by enclosing them in it for several days, by means of a slight lath railing, giving them an abundant supply of food and water during their imprisonment. Birds like to select their own partners.

DISEASES.

The MEGRIMS is an incurable disorder, in which the pigeon moves about and flutters at random, with its head turned in such a manner that its bill rests upon its back.

If the birds suffer much while MOULTING, the best remedies which can be adopted are to keep them in a warm place, put plenty of hemp-seed in their ordinary food, and likewise saffron in their water.

When the birds are affected with the WET ROUP, give them a few pepper-corns once in three or four days, and put a handful of green rue in their water.

The DRY ROUP is a kind of dry, husky cough: it arises from a cold, and to cure it, administer three or four cloves of garlic every day to the little patients.

When your pigeons are infested with INSECTS, you will find that smoking their feathers thoroughly with tobacco is the best curative.

The CANKER is occasioned by the cocks pecking each other, which, as they are extremely irritable, they often do; and to cure it, the parts affected must be rubbed every day with burnt alum and honey.

When the incrusted flesh round the eyes of carriers, barbs, or horsemen is injured or pecked, it should be bathed with salt water for several days, and if this remedy does not succeed, another composed of two drachms and a half of alum dissolved in an ounce and a half of water should be tried.

When pouters and croppers have fasted somewhat longer than

ordinary, they are apt to gorge themselves, and when this occurs, put the bird, feet downwards, into a tight stocking, smoothing up the crop, so that overloaded as it is, it may be kept from hanging down; then hitch up the stocking on a nail, and keep the bird a prisoner until its food is digested, supplying it with a small quantity of water now and then. When the bird is taken out of the stocking, it should be put into an open coop or basket and fed scantily for some little while.

If the birds are lame, or the balls of their feet become swelled, whether from cold, or from being cut with glass, or any accident, the most effectual remedy is a small quantity of Venice turpentine spread on a piece of brown paper, applied to the aggrieved part.

LAWS RELATING TO PIGEONS.

By Act of Parliament of 7 and 8 Geo. IV. c. 27, it is enacted, that if any person shall unlawfully and wilfully kill, wound, or take any house-dove or pigeon, under circumstances not amounting to larceny, upon being convicted thereof before a justice, he shall forfeit over and above the value of the bird any sum not exceeding forty shillings. But it has nevertheless been determined that the owner of land may kill such pigeons as he may find devastating his corn.

CONCLUSION.

As many pigeon-dealers are in the habit of playing off innumerable tricks upon youthful, inexperienced fanciers, it is highly necessary to have the advice of some person well acquainted with pigeons, when making a purchase, especially if a select stock is required.

It is a good plan, before having a flight of pigeons, or at least of letting them leave the loft, to know what character the neighbouring pigeon-keepers bear, for the fancier must never expect to have a good flight if his neighbours are adepts in the art of enticing birds into their own aëries. True enough, he may do the same with theirs if he has the chance; for according to the old saying, "what is sauce for the goose is sauce for the gander."

PET ANIMALS.

THE DOG.

In prosperity, in adversity, alike true to his master, come weal, come woe, is that faithful friend and engaging companion—the dog. He is as true to the beggar boy in his tatters and his hunger, as he is to the gentlemanly youth, who can feed him on the fat of the land. Let him but find a kind master, and nothing but death will divide this faithful animal from him; nay, he has been known to lie days and nights on the grave of his dead master, and never more to be persuaded to taste food; but through rain and cold, light and darkness, remain there until he died. The love of father or mother could not go beyond this, it is the "utmost bound of sorrow." What weary miles he will follow his master, no murmur, no complaint; a kind word, and there is that warm, old, pleasing wag of the tail, and he has his feet on your knee in a moment. Was there ever a boy born that did not love a dog? Then he is so full of play, too, quite as fond of it as any boy can be; so that the boy who has a dog has always got a playfellow, and a good-tempered one too, who will not sulk and quarrel if a cross word is spoken to him. And how fond he is of children; we hardly ever remember a thorough-bred dog biting a child; some cross between a dirty-bred-kennel-and dust-bin one might; a mangy, surly cur, but never your fine clean-built, bright-eyed, thorough-bred dog.

Then if they see you put your hat or cap on, to go out for a walk, how they bark and bound with delight, as if to say that they shall enjoy the ramble quite as much as you will. Then, what a hunt he will have in every hedge and ditch; you hear a water-rat go "plop" here, and see a weasel running off there, and when they escape he comes barking to tell you, as well as he can, how annoyed he is that he couldn't catch one of them to place at your feet. Then he is so faithful, too, if well trained; leave him what you will to guard, and tell him to "mind it," and whether only an old rag, or a purse of gold, he will only part from it by sacrificing his life, until he gives it up to his master. To write the history of the dog, and tell all we know and have heard of him, would be to fill this volume, and then we should not have said half enough about him. He can do almost everything but talk; and there are instances on record where he has almost done that—where he has whined and dragged at the skirts of his master's coat, to draw his attention either to something that had happened, or would have happened but for his timely discovery.

Dogs have done things that have shown retentive memory, forethought, calculation—nay, even reason; for instinct is a poor word to use to express what some dogs have done. We have known a dog remain in a room for weeks together with a sickly boy, and never quit it beyond a minute or two at a time.

About the true origin of the dog we know nothing. At what period of time he ran wild, and when he first became domesticated, history tells us not; but so far as the most ancient records go back, we find him the faithful attendant on man; in the oldest Eygptian scrolls that have been discovered, he is pictured as standing beside his master. A great man, after much patient investigation, came to the conclusion that the wolf, jackal, and dog are all of one species. When dogs return to a wild state they will, after three or four generations, bear a close resemblance to the wolf, while a wild dog domesticated, however wolf-like he may at first appear, will, after a few generations, assume the distinct marks of some familiar breed, and lose all his wolfish appearance.

THE BLOODHOUND, used for tracing thieves, murderers, deer-stealers, wherever a foot-print was left or a drop of blood shed, is an old type of dog that no doubt has come down to us through long centuries unaltered. His scent is very acute. His colour is generally black and tan, and he is a large, powerful dog, often standing nearly three feet high, and weighing as much as eighty pounds; his pace is rather slow.

THE WOLF-DOG is all but extinct. He stood nearly four feet high, and could master a wolf single-handed, in shape he resembled the greyhound, but was a rougher and a stronger dog. There were one or two remaining in Ireland some years ago. He belonged to the olden time, and was no doubt often used by the ancient Britons when they were compelled to pay the tribute of wolves' heads to their Saxon conquerors.

THE DEERHOUND is another old dog often mentioned in our ancient chronicles and early ballad lore; he belongs to the class of greyhounds, and the latter are too well known to require a single word of description.

THE GENUINE BULL-DOG is a terrible fellow to look at, and has been known to turn on his master when offended. But he does not worry and shake his victim like the terrier or the mastiff, nor when he once gets hold will he leave go unless taken off by his master, or compelled to leave go by main force.

THE MASTIFF, as his under-jaw tells, has a cross of the bull-dog about him; he is a powerful dog, and makes a capital guardian.

THE SHEPHERD'S AND DROVER'S DOGS are wonderfully sagacious, and the marvellous stories told of the wondrous things they have done, are enough to place them in the highest class of intelligent dogs. Many a flock of sheep has been saved by their sagacity and perseverance.

POINTERS, SETTERS, SPANIELS, RETRIEVERS, &c., are sporting dogs; though all very useful and faithful to their masters, they are not, with an exception or two, such dogs as boys would select as companions, though we could find a good deal to say about them, were we writing a history of dogs, instead of merely skimming over their different varieties to arrive at such as we do consider companiable, and these are the English and Scotch terriers.

A TERRIER is the dog a boy ought to have; he will fetch and carry, hunt vermin, guard the house, bark when you bid him, or be mute in a moment—a terrier is the dog for a boy, after all. Oh, what a fellow he is for a rat; won't he shake him and make him squeak; I wouldn't be a rat in the jaws of a good terrier, if I might be made the Emperor of China to-morrow. Or let him get hold of a stoat or a weasel, and he'll just show them as much mercy as they show a poor little rabbit or hare. It's all up with them, I can tell you; and if they look into his eyes, they'll find no more pity there than they would in a pebble. If he can't worry a hedgehog, he'll make it "shake in its shoes," and even have a turn at a badger, which is as sharp and close a biter as a bull-dog. Then the Scotch terrier seems as fond of water as a fish: throw your stick in and he dashes after it, comes paddling up with it in his mouth, and places it at your feet. To say nothing of their playfulness, they are such faithful animals too, and would follow their master to the end of the world, and further, if he went there. Then they may be taught no end of amusing tricks—to beg, to walk on their hinder legs, to sit up until you count any given number, to fetch anything you have dropped purposely, and left a long while behind, if it has been pointed out to them. And many a laughable anecdote is on record, of their masters having thrown things away, which they had no wish should be found in their possession, when, lo! to their confusion, the dog would come up with the rejected article in his mouth, and place it at his master's feet, when the latter would have given all he possessed to have been ten miles away another road at the moment. We have heard of a man stealing a small bundle, throwing it away when pursued, of the dog picking it up, and the thief being captured, through the pursuers following the footsteps of the faithful dog. Another man went to purchase a flock of sheep, which his dog drove into a corner for him to examine, but as he could not agree with the owner of the sheep about the price, he went away without them. Not so the dog; he had been set to drive them in a corner by his master, and knowing nothing about the disagreement respecting price, why he drove them out of the field to his master's farm-house, which was ten long miles off. The dog's master was tried for stealing the sheep, and had a very narrow escape. Two tollgate keepers proved that the dog barked in the night until they opened their gates and let the sheep through, thinking the master was behind, but they found he was not, and this saved the man's life, for sheep-stealing was punished with hanging in those days.

Now, in keeping dogs, if they have kennels, always clean them out once a week, and be sure and take care that they never want for water. Never leave anything hot in their way, for if scalding, and they are ravenous, they are likely enough to eat or lap it, though it kills them. For the mange, rub the spots with sulphur ointment, and this, unless it is a very virulent kind of mange, will cure him in four or five days; tumours which consist of small soft bladders of a circular form, lying close under the skin, must be cut out with a sharp knife; this, if skilfully done, gives the dog but little pain, and if the skin is again drawn carefully over, the wound will heal of itself. No boy need be told what a dog will eat; you may safely give him anything from a full grown elephant to a whale, and with time enough, he will finish them both.

THE SQUIRREL.

This is one of the most beautiful animals that can be tamed and kept within doors, and we know nothing beside that has so splendid and handsome a tail; for altogether it is a perfect little beauty, and of such a rich red-brown colour as is quite delightful to look at. The squirrel is a native of our woods, and went skipping about among the branches, just as he does now, when the bearded old Druids worshipped the oak. There is no animal in England can leap such an immense distance as a squirrel, for it has been seen to spring from one tree to another when the broad, wide, common highroad ran between the two trees. Should it happen to fall on the ground while making these terrific springs, it alights on its feet, and is off and up at the top of the next high tree quicker than the eye can follow it. It is a pretty sight to see it sitting and cracking a nut, which it holds in its

PET ANIMALS.

fore-paws, that are like hands, with which it carries everything to its mouth, while the beautiful tail curls over its head, like a plume in a cuirassier's helmet. In a wild state it lives upon nuts, acorns, leaf-buds, beechmast, the bark of certain trees, and even the young tender growing branches. The nest, or dray, as it is called, is generally built in the hollow of some old tree, similar to what the woodpecker selects, and is formed of moss, twigs, and dry leaves. It breeds in May or June, has four or five young ones at a time, and there is nothing prettier to be found in our old English woods than a nest of young squirrels.

Though naturally a wild and timid animal, when caught young it soon becomes so completely familiarized to a state of domestication, that it may be suffered to run loose about the room, or even the garden, without making the slightest effort, or evincing the least inclination to return to its natural wild state. After gambolling and frolicking about for a while, it will come back when called by name to its protector (and it should have a name to answer to, Bob is as good as any), and crawl over him, nestle in his bosom, and display other signs of the strongest attachment. When taken and bred from the nest, it may be taught to perform a variety of amusing little tricks, such as to dive into the pockets for a nut; to run after one thrown along the ground, like a dog; to turn over head-and-heels; to jump over a stick, or from one hand to the other, when held a considerable distance apart; besides many others. Indeed, it has all the docility, as well as the affectionate disposition, of a dog; it will recognise as soon the person of his master, and displays the same fondness and attachment to him.

There are various sorts of cages made for these little animals, varying in price from five shillings to two or three sovereigns, or even more. The one most generally in use is the circular-topped cage, represented in the cut. But it is cruel to put so beautiful a creature into such a cramped-up prison, so fond of liberty as it is in its natural state. It has a little sleeping-box, which opens with a lid, so that the bed may be changed, and the place cleaned out without difficulty; this communicates with  the open cage by means of a hole, just large enough to admit the body of the animal, furnished with a sliding door, so as to be stopped up or left open at pleasure. The outer cage is fitted up with a sliding bottom and small tin trough for food; the edges of the woodwork must all be carefully covered with tin, otherwise the little tenant, by continual gnawing, would not only greatly disfigure the cage, but eventually effect his escape. The revolving or turn-about cage is another often used, but this is decidedly objectionable; the motion is an unnatural one, and must therefore subject the poor little prisoner, until he becomes accustomed to it, to a great deal of unnecessary torment. They are also sometimes kept fastened by means of a small brass collar and chain, like a dog, to a little box or kennel, with a platform in front; they must, however, be thoroughly tamed before they can be kept thus, otherwise by their continual efforts to escape they are liable to strangle themselves; but the best way after all, when they are thoroughly tamed, is to let them have the range of the room, or to purchase a cage six feet long and four feet high; in this there will be room enough, and it ought to have perches like the branches of trees. The cage should be cleaned out regularly every day, to prevent its getting offensive, and a little gravel sprinkled on the bottom. The sleeping-box should be furnished with some sweet hay, moss, &c., with a little wool, about the breeding time.

They should be kept on the same sort of food they obtain when in their natural state; bread and milk may also be added, but it must be perfectly fresh and sweet. They may be purchased of most bird-fanciers for about four or five shillings; when very tame higher prices are asked, sometimes as much as fifteen shillings or a sovereign. Numbers are also brought up to London by country market people, and sold about the streets at more moderate charges.

In purchasing, care should be taken to select a young one, which can be readily distinguished by its beautiful white teeth, for when the animal is old the teeth are yellow.

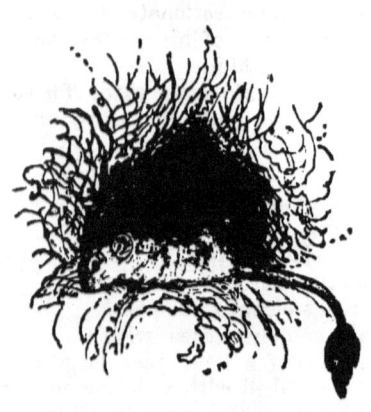

THE DORMOUSE.

THIS is a very clean, handsome little animal, whatever people may say to the contrary, and is a great pet with boys in the country; it has also a fine long bushy tail, which, when sitting down to wash itself with its pretty hands, curls over its back like the squirrel's, though we are not going to say it is so beautiful as the squirrel. It lives in copses or woods, and is so partial to company, that ten or a dozen nests have been found close to one another. Its motions are so quick that it cannot easily be taken; you just get sight of it, and it is off like a shot, in an instant.

In size it is rather larger than the common mouse; the colour is a light brownish red, inclining to white on the underneath part, particularly on the throat, where it is almost a pure white; while the eyes are large and black.

It is a dull, sluggish little animal, but perfectly harmless and inoffensive in its temper, displaying neither aversion to those who annoy, nor much affection to those who caress it. It is by no means shy, soon becomes domesticated, and will suffer itself to be handled freely, without evincing the slightest marks of displeasure. Like most of the mouse tribe, it has four front teeth, but it seldom makes use of them as a means of defence. It builds its nest with grass, moss, dry leaves, &c., in the very thickest parts of some strong quickset hedge, generally near the bottom where the root divides off into a number of small upright branches; here it so resembles the clump of dry leaves usually collected there, as to escape being noticed except on the strictest search; even when discovered it is so completely protected by the cluster of stems, that unless these are cut through, it is a matter of no small difficulty to secure it, without doing injury to the little ones, should any happen to be inside. It feeds on all sorts of fruit, acorns, beans, peas, and even corn. During the winter it remains in a state of torpor, in preparation for which, towards the end of autumn, it makes itself a small round compact nest of the sharp prickly leaves of the fir tree, just large enough to contain its body; into this it creeps on the approach of cold weather, and remains tightly coiled up in the form of a ball, till reanimated by the warmth of spring. An occasional fine day, however, in winter, will revive it, and tempt it from its retreat; but the returning cold quickly compels it to resume its former state. Moderate warmth will always restore it to life, but sudden exposure to the heat of fire destroys it; but a gentle warmth, such as holding it in the hand, rouses it up without injury. Whilst in this state of torpor, it appears to possess but little more warmth than any inanimate object, and to be deprived of all susceptibility to pain, except that of a very acute description; it may be rolled or even thrown about without being roused in the least from its deathlike lethargy.

They breed but once a year, and bring forth four or five young ones at a litter.

They are usually kept in pairs, in cages similar to those used for the squirrel, only on a smaller scale. They may be fed on all sorts of fruit, especially of the nut kind; a little bread and milk may also be given them, of course fresh every day. Their cages must be kept perfectly clean; the sleeping-box should be furnished with soft moss or hay, to which a little cotton may be added when they are about to litter. The male is seldom known to devour the little ones, and, therefore, need never be separated from the female. During winter, they should be kept in a warm room, when they will remain in an active state all the year round, like most other animals.

They may be bought of bird-fanciers for about eighteen-pence or two shillings a pair. In the winter time, when the cold has driven them to their nests, they may be found in great numbers in the hedge-rows; the bushes being then stripped of their foliage, their little hiding-places are exposed more readily to view

WHITE MICE AND HARVEST MICE.

THERE are few lads that have not, during some portion or other of their school days, amused themselves by keeping these pretty little creatures. The moderate sum for which they may be bought, and the trifling expense incurred in keeping them, place them within reach of almost every schoolboy's pocket; while the constant amusement they afford by their innocent antics, and the tameness and docility of their nature, amply compensate for the little trouble they require at his hand. So well must they be known to all our young readers who possess the slightest love for natural history, that a long account of them is altogether unnecessary; a few remarks, therefore, on the mode of treatment, &c., is all that need be offered.

They may be purchased at all bird-shops; eightpence or a shilling are the prices asked for a pair of young ones; after having had a litter, eighteen pence to two shillings; the same price is also asked for a doe with four or five little ones. They are exceedingly prolific; six or eight broods are frequently produced by one doe in the course of a year, and from three to eight young ones at each birth. When the female is about to litter, the buck is frequently separated from her, and kept in another cage till the young ones are a week or ten days old, lest he should devour them. Though this circumstance does occasionally happen, yet it is by no means a common occurrence; if, however, he has once done it, he will be always likely to repeat it. Two or three kinds of spiders are quite as hard-hearted, and the wives think nothing of gobbling up their husbands, "body and bones," if they once fairly get hold of them.

The cages for white mice are as varied in price as they are in form, material, and neatness of execution. A common one may be had for a shilling, or even sixpence, and more finished ones at all intermediate prices between that and a sovereign. The usual and most convenient cage is that like the squirrel's, on a small scale; the same as recommended for the dormouse. Some are very ingeniously contrived like little houses fitted up with glass windows, doors, &c.; they consist frequently of two or three stories communicating one with another, by means of small ladders, up which the little animals are compelled to climb to obtain their food, which is generally placed in the uppermost story; others are made like small models of windmills, the sails of which are made to turn round by means of a revolving cage inside. The common revolving or "turnabout" cage is one often used; but the same objection is to be raised against it for these little creatures as that mentioned before with regard to the squirrel. The usual food for white mice is bread sopped in milk, squeezed pretty dry; oats, too, they are very fond of; a few may be given them every day. Cheese, of any kind whatever, is decidedly bad.

Great attention should be paid to the cleaning out of the cage, the health of the little tenants mainly depending upon it; it should be done regularly every morning, and that thoroughly; the smell, which is always very strong, will otherwise be exceedingly offensive; the bed, too, should be frequently attended to and changed; after littering, however, the sleeping-box should not be opened at all for three or four days. The young ones are able to shift for themselves when about a fortnight or three weeks old.

A very pretty piebald may be obtained by a union of the white with the common brown mouse. There are besides several other varieties of these pretty little animals, such as the black, the black and white, black and brown, fawn-coloured, &c.; but all these being scarcer than the plain white, are much more expensive, a pair of them costing as much as four, five, or even six shillings.

Harvest mice and the long and short-tailed field mice are also frequently kept by boys as pets, particularly by those in the country. They are found about harvest time in great numbers on the cornfields; their nests also are found in the hay-fields after the grass has been mown. They soon become tame and familiarized to confinement, and have none of the offensive smell the common species have. They may be kept on oats, beans, peas, nuts, &c., with a little bread and milk.

The harvest mouse, perhaps, is the most interesting little animal. It is the smallest quadruped known in the world, being only one-sixth the size of the common house-mouse. It is nothing near so big as the large bees you often see humming amongst the clover in summer. Two harvest-mice full-grown will barely weigh down a halfpenny. It generally builds its nest near the top of three or four ears of corn, the plumed ears making quite a screen overhead; the stalks or stems of corn are quite strong enough to support this little nest, which it fastens to the straw by twining the grass round the stems. The nest, with its eight or ten young ones, is hardly so large as a regular-sized cricket-ball. It will amuse itself in a wheel-cage, like a squirrel, and seems to enjoy itself very much; it will also twine its tail round the wires like a monkey, and swing its tiny, inch-long body to and fro like a pendulum, suspended only by its tail.

THE GUINEA-PIG.

The Guinea-pig, or restless cavy, is a prettily-marked, stupid little animal, which came originally from South America, and has long been a favourite with most little boys, for when a boy becomes a youth, he aspires to keeping something more interesting than these senseless little squeakers. Still it is a pretty sight to see the old ones followed by two or three litters, each lot not more than two months older than the last addition, and to watch their antics when pleased, which consists of a squeak and a peculiar sharp turn, as if they tried to jump out of their skins, but couldn't, as they are fitted in too tightly for that. As for the use they are, why, they eat and sleep—

> "Just do nothing all the day,
> And soundly sleep the night away."

It is, however, a pretty, harmless little animal, but as a pet is far inferior to others that are generally kept. Though gentle and inoffensive in its manners, it seems incapable of feeling the slightest attachment for those who feed and caress it—even for its own offspring it evinces little or no affection; it will not only suffer them to be destroyed before its face, without making the smallest efforts to defend them, but will even at times devour them itself. For all useful purposes they are utterly valueless, though their flesh is used as an article of food in their native country, but their skins, notwithstanding the beautiful sleekness of their appearance, have as yet been turned to no account by the furrier. Their only recommendations, therefore, are the gentleness of their dispositions, the cleanliness of their habits, and the beautiful colouring of their coats. In this latter respect they are very varied, black, white, bright reddish brown, and mixtures of the three, called tortoiseshell, being the principal varieties; the latter are the most prized, particularly where the dark colours predominate.

In their native country they are generally of a pure white, with pink eyes, and it is nothing unusual to have one out of a litter white with pink eyes in this country. The alteration which has taken place in them in this respect is perhaps to be attributed to change of climate, food, &c., an improvement that is to be seen in most animals which have been domesticated by man.

They possess amazing fecundity—to a greater degree, perhaps, than any other four-footed animal. They bring forth six or eight times in the course of a year, and from four to twelve young ones at a litter, beginning at the age of two months. The average number which one female is the means of producing in one year is estimated at six hundred; thus, in a short time they would increase to such an extent as to set computation at defiance, were there no check to the multiplication of the species.

Rats are supposed by many people to have a great antipathy to guinea-pigs, carefully avoiding the place where they are confined. Under this impression, which, however, is an erroneous one, they are frequently kept by fanciers in their rabbit-houses and pigeon lofts, as a means of protecting their stock against the depredations of these rapacious vermin. They are allowed to run almost anywhere, and to shift for themselves; no attention whatever need be paid to the feeding of them, the mere refuse scattered about the floor being sufficient for their subsistence.

When, however, they are kept for amusement, their cages are generally made precisely similar to the rabbit's hutch, only of rather smaller dimensions; their treatment too, in most respects, is much the same as that pursued with regard to those animals. Their ordinary food should be oats given twice a day, and not too many at a time; they are also very fond of bran, which is a cheap diet, and they will fatten upon it, if allowed plenty of exercise, and keep healthy. Green meat should also form a portion of their usual diet, particularly the wild sorts, such as dandelions, sow-thistle, plantain, &c.; tea-leaves they are remarkably fond of, but these should only be given them now and then by way of a treat; bread also they are very partial to, dipped in milk or water.

They are sold by all bird-fanciers; the prices varying from sixpence to half-a-crown, according to their age, colour, &c.; as before remarked, the dark rich coloured tortoiseshell ones are considered the most valuable.

THE HEDGEHOG.

THIS, though a rough and prickly customer to handle, is a clean little animal, and a great pet with country boys. If you have ever seen one—and they are commonly sold at the shops kept by men who call themselves bird-fanciers, or even hawked about the streets of London by countrymen—if you have seen one you are aware that, saving the belly, they are covered with sharp spikes, and that when alarmed, or whenever they please, they have the power of rolling themselves up into a tight round ball, which shows nothing but spikes, and may be rolled along like a cricket-ball, without causing the little animal to uncoil itself, while every spike is firm and erect as a needle, and almost as sharp. While in this shape very few dogs can worry the hedgehog, and as for the fox, who is rather partial to him, true to his wily nature, he rolls the poor hedgehog along with his paws until he comes to a pool of water, when the hedgehog unrolls himself, exclaiming, no doubt, "Hey, what the deuce is this; why I can't breathe!" and trying to peep about and see what's the matter, the fox, on the look out, seizes him by the belly and eats him all up saving the spines. Were you to thrust a dozen pins with large heads through a piece of parchment, you would have an exact representation of the spines of the hedgehog, every one of which is retained inside the skin by the large pin-like head.

Some say it sucks the cows and draws off their milk, but this is absurd, as its mouth is not adapted for sucking: its favourite food is insects and snails; it also feeds on frogs and mice, and will even kill a snake and eat it all up, beginning at the tail. The way it kills a snake is very curious: it gives the snake a bite on the back, then rolls itself up like a ball, remaining still as a stone for some time, while the snake lashes and writhes about in agony; as soon as the snake is a little quiet the hedgehog gives him another sharp bite on the spine, and so continues until the snake is killed, then he begins at the tail, as the Rev. J. G. Wood tells us in his beautifully "Illustrated Natural History," and eats him up "as one would a radish." In a natural state it sleeps all the winter, rolled up in a hole which it has filled with grass, moss, or leaves; and when domesticated it will hide itself in some dark corner for weeks, and never once make its appearance unless it chances to awake and feel hungry, then some day you will see it come creeping towards the fire, and be very glad to see it too. They are great destroyers of beetles, eating them up as you would a handful of raisins, and seeming equally fond of them. They need no looking after at all, but will take care of themselves, though it is as well to have a little hutch to put them into now and then. Their feeding time is in the night, and if there are black beetles in the kitchen the best plan is to leave the hedgehog there, and let him devour all he can catch. They have four or five young ones at a litter, which are born blind. It is no uncommon sight to see a countryman with both old and young ones to sell. You can buy a young hedgehog for sixpence, and an old one for a shilling, or less than that even. We know of no animal that is less trouble to keep, and in time it becomes so tame as to come out of its hiding-place when called.

RABBITS.

Perhaps of all pet animals—always excepting the dog, who is the companion and faithful friend of both men and boys—there are none that are greater favourites than rabbits; they are such pretty things, with their long ears and diversified colours, and may be kept so clean, too, if well looked after. Then they find a boy plenty of occupation when he takes a country walk, or even if only in the suburbs of a city, for where does not the dandelion and sow-thistle grow? and there is no green meat that rabbits are fonder of than these wild plants. Building the rabbit-cote, too, is pleasant employment; buying the tea-chest for the hutch, too, and making it, finds a boy something to do, then nailing laths before it, making a door with leather hinges, then a trough for the rabbits to eat out of, all of which require some skill, and which a clever lad will do.

They are such playful things, too, if they have plenty of room; and happy is the lad who can get some little outhouse allotted to him where he and his rabbits can have it all to themselves. How they will frisk, and leap, and play with one another, and keep so healthy with such exercise, that not a pot-bellied one will be found amongst them.

Some boys are satisfied with keeping common rabbits, whilst others, more chary in their taste and choice, stock their little rabbitries with the more expensive fancy ones, which, in point of appearance and beauty of colour, are unquestionably more worthy of attention. It is not essentially requisite that the young fancier should procure first-rate animals when he first lays in his stock, as he may purchase very good rabbits deficient in one property or the other, but which will produce, as frequently as the best, first-rate young ones, for a comparatively moderate price, often at much less than a fourth of what would be asked for them were they perfect in all their properties. Fancy rabbits are rather more delicate, and require a little more care in their management, than the common ones; but as they want neither more nor better food, the extra trouble is worthily bestowed.

THE WILD RABBIT.

The wild rabbit, if not a native of this country, was known in England at a very early period, for we find one of the favourite diversions of the people in the early ages was to let these animals loose amongst crowds of spectators for boys to hunt; the interest of the sport being the confusion caused by the endeavours of the poor affrighted little creatures to escape amongst the bystanders, and the hearty and boisterous glee with which the juvenile huntsmen followed and strove to secure their timid prey. The methods then in use for snaring and taking them alive were similar to those employed at the present day.

Though the wild rabbit in its general appearance greatly resembles the hare, yet the two species never intermix or inhabit the same tract of country, the rabbit dwelling in a burrow or hole, called a warren, where large numbers congregate together, which it makes for itself in banks or broken ground, generally in a district where the soil is of a sandy or gravelly nature; whilst the hare chooses its retreat near some low bush, fern, or other slight shelter, on rich and somewhat flat and dry ground. Rabbits are destroyed in immense numbers by various methods, of which ferreting is one of the principal; the ferret is muzzled, and having a bell fastened round its neck, is turned into one of the chief holes, which is then, with all the surrounding ones, carefully covered over with little purse-shaped nets, firmly secured by means of wooden pegs; the rabbits, terrified by the appearance of their natural enemy, immediately rush to one of the openings, where they become entangled in the net, and are caught. Hunting them with dogs is another method of destroying rabbits; terriers and spaniels are generally used for this sport, which is mostly carried on in the autumn, when the crops have been gathered in, and as the little animals frequently lie at that period of the year in hedges, often at some distance from a burrow, they stand but little chance of escape from the attacks of two or three active dogs. Great numbers are also shot.

The wild rabbit is amazingly prolific, breeding four or five times a year, and producing from five to seven, eight, or nine even, each time; but from its numerous four-footed and flying enemies, such as weasels, foxes, polecats, falcons, kites, &c., the effects of damp, and the singularly-unnatural propensity of the old bucks to eat up the little ones, the race is prevented from increasing to an obnoxious extent. The most beautiful variety of the wild rabbit is that brought from the vicinity of the city of Angora, in Asia Minor, and from thence called the Angora rabbit; it is an extremely pretty animal, being covered with long silky hair or fur, which when dressed forms a valuable article of commerce.

THE TAME RABBIT

Was originally the wild rabbit which we have domesticated, and made larger by feeding and attendance, and which soon assumes its wild shape and habits again if allowed to escape. Common rabbits vary exceedingly with regard to colour, some being entirely black, others white with red eyes, others mouse-colour, others fawn, some brown, and some grey with tawny feet. Persons who are particular with respect to the colours of their rabbits, should endeavour to ascertain the colours of the does from which their stock came, for it often happens that rabbits produce litters in which not one young one of their own tint can be found; for instance, if a cross of grey happened to be in the stock some four or five generations back, it may appear again, although all your breeding rabbits are of other colours. Grey is the worst of all colours, in the opinion of the fancy, and the most difficult to get rid of, yet it does not always happen that grey rabbits throw litters of their own colour.

When choosing does for rearing, take the largest from those rabbits which have the fewest in their litters, as it is supposed that when the does have but few at a time, the young ones are more likely to turn out fine; let the little ones remain with their mothers until they are about six weeks old, then take them away, and keep them in hutches, two together, for about the same period, and as they become excessively quarrelsome in their dispositions when near four months old, they must then be separated. When you lift up your young rabbits, always take hold of them by the ears, and place one hand under the lower part of their backs, for it is injurious to handle them too much.

Although does will breed at the age of six months, it is better that they and the bucks be ten or twelve months old before they are first put together; the rabbits should not be left together for more than ten minutes. The doe goes with young thirty days, and towards the time when she may be expected to kindle, fresh hay or oat-straw, or both, should be given to her for a bed; when she nibbles the hay or straw into little bits, it is an unfailing proof that she is with young ; and a few days before kindling, she tears the soft flue or fur from her body, to make a nest for her young ones. If, as it often happens, some does have a large number of young ones at a kindle, and others but few, it is as well to equalize the number for each to rear, by taking from those which have the most and giving them to those with the fewest.

About six weeks after kindling, the old rabbits may be put together again, but if the doe has had a large number of sucklings, a longer period should elapse. If the doe is weak after kindling, a malt mash, made of fine pollard scalded, or barley-meal with a small quantity of cordial horse-ball mixed up with it, will be found beneficial. Bread soaked in milk and then squeezed rather dry, will also strengthen her materially, if she can be made to take it. Those fanciers who keep their rabbits mewed up in hutches, should not let their does have more than three litters in a year. It is necessary to protect the young rabbits from the old bucks, otherwise they will be certainly devoured; and as rats and other vermin are particularly fond of such delicate morsels, making a meal off all which come in their way, it is requisite to construct the hutches so as to keep them out. If you have a doe which possesses so little affection as to destroy her young ones, fatten her at once, for the sooner she is killed and eaten the better. Pink-eyed ones they say are the most guilty.

To the feeding of his rabbits, the young fancier must pay great attention, he should carefully see to them twice a day, at the very least; that is, early in the morning and in the evening, and according to the rule of many fanciers, give them another meal in the middle of the day. Abstain from giving a superfluity of food, else they will become cloyed, and waste also what they cannot eat. The most suitable food consists of the delicate tops of carrots, celery, parsnips, hare parsley, and furze; the leaves and roots of white beet, stalks of dandelions, sow-thistles, and lettuces; grass, clover, tares, coleworts, and cabbages in moderation, apples, pears, pulse, corn, and Jerusalem artichokes. Cabbage leaves should be given with discretion, as they are apt to disagree with the animals; in fact, too much green food is injurious, being likely to produce a disorder termed *pot-belly*. To guard against an evil of that kind, a due proportion of dry food, such as fine fresh hay, pea-straw, or corn, should be frequently given with the moist vegetables; it is very different with wild rabbits that have plenty of fresh air and exercise. When it is impossible to procure greens, roots, or grains, the corn may be slightly moistened with water or milk, and indeed, during a dearth of fresh vegetables, a small quantity of tea-leaves, squeezed tolerably dry, will be found to agree very well with the rabbits. Some fanciers give a tablespoonful of water, beer, or milk occasionally to their rabbits, when corn forms the chief part of their aliment, but never when greens can be obtained. Rabbits accustomed to live chiefly upon bran or any other kind of dry food, will eat with great avidity the parings of turnips, apples, or pears. Potatoes, either boiled or roasted, may be given amongst other food, but never when raw, as they are then unwholesome. When a doe has a litter by her side, it is a good plan to soak the split or whole grey peas for a few hours before they are put into the trough, and if peas are given to recently-weaned rabbits, they should be also soaked. Although we recommend the food to be given in such quantities only as the animals can eat in a few hours, yet when a doe is about to litter, she may have somewhat more allowed her; when she suckles, also, a greater quantity of food must be given to her, as she will then eat twice as much as at other times. As soon as the young ones begin to nibble, they must be supplied with a liberal allowance of food, three times a day, punctually. If the aim is to fatten rabbits for the table, the best age to put them up for that purpose, is from five to eight months, and the kind of food most suitable (and on this subject various opinions are held) is barley-meal, oatmeal, or split peas, or a mixture of them, with the addition of a little sweet hay. A tablespoonful of water a day may be added, and a small quantity of carrot-tops, sweet marjoram, parsley, and basil, may also be given daily, with advantage. The more the food of those put up for fattening is varied, the sooner the end will be attained; but when the animals are once full-fat, as the breeders express it, they often pine away and lose their plumpness. Practice, alone, will regulate the exact quantity of food each rabbit should be allowed, as no precise rules can be laid down for the purpose. If they can be allowed to disport in a yard for an hour or two, in fine weather, it will add much to their general condition and health; of course, too much freedom and exercise must not be permitted, as that would militate against their speedily becoming plump and fit for table.

The most careful and regular attention should be paid to the feeding of all pet animals, for it is very cruel to neglect those poor little things, whose existence depends upon the supply of food afforded to them, and which if perishing from lack of nourishment, cannot escape from their captivity to seek a kinder home elsewhere.

FANCY RABBITS.

The fancy in rabbits is very changeable; some years ago, a fine common rabbit of two colours was esteemed a fancy one, but now, a rabbit must possess certain properties, many of which are never found in the common kinds, before it can be classed as a fancy specimen; these properties consist in a perfectly symmetrical shape, good arrangement of colours, full dew-lap, and a peculiar position of the ears. In addition to a perfect shape, a rabbit must have what is termed a "good carriage"—that is, its back should be finely arched, and its head held so low, that its muzzle and the tips of its ears may almost touch the ground. Many fancy rabbits have their fore-legs bent inwards, but this, although it appears a deformity, is not considered of any importance, neither does it lessen the value.

The correct arrangement of colours is a very important point, and rabbits are divided into three varieties, distinguished by the colours of their fur: these are, the lead, or as it is technically called, the blue-coloured and white, the black and white, and the tortoiseshell; and these varieties are again subdivided into three classes, from the peculiar arrangement of the spots of colour on their faces, termed the single, double, and butterfly smuts: of these, the latter is the most valuable. The single smut is a solitary patch of a dark colour on one side of the nose; the double smut, a spot on each side; and the butterfly smut, a spot on each side of the nose, with a dash of colour on the nose, forming altogether a slight resemblance to a butterfly. Hence the name. If a black and white rabbit's face is ornamented in this manner, it is said to be a black butterfly smut; and if a lead-coloured rabbit shows this mark, it is called a blue butterfly smut. It is not really necessary that a fancy rabbit should possess these markings, but if it does, its value is enhanced. Other marks must likewise be well-defined upon the rabbit, before it can be fully esteemed a perfect fancy one; thus, a patch of dark colour should be on its back, this is termed the saddle; its tail must be dark, with dark stripes also on each side of its body, in front, which from their passing backwards so as to meet the saddle and, as it were, form a collar, are styled by the fanciers the chain: the animal's throat may be mottled with dark colour and white, but its legs and belly must be of the most perfect snowy whiteness. Neither must the spots of colour be grizzled, nor have many white hairs amongst them, for if they have, the beauty and wholeness of the animal's colour are much diminished; neither should the saddle terminate in a harsh, abrupt manner, but have its edges broken by dark spots, lessening gradually in size, and ending with the chain on the shoulder; these spots, of course, must also be free from white hairs. It very seldom happens that rabbits exactly perfect in point of colour can be procured, perhaps scarcely one in a hundred; the nearer they are to the rules, however, the more they are valued—at least in as far as the property of colour is concerned. It also sometimes happens that very good does produce young ones which are merely touched with dark colour—that is, with only a spot or two round the eyes and on the back, and perhaps a dark nose; these generally are weakly animals.

The dewlap is a property peculiar to fancy breeds, and is a highly prized one, on account of the noble effect it has; it is a protuberance formed of skin and fat such as you are familiar with through seeing oxen, and is not developed until the rabbit has nearly attained its full size; it commences immediately under the jaw, goes down in front of the chest, and ends between the fore legs; it is indented in the middle, and is frequently so large, that when the animal is in a state of repose, with its head drooping, it projects on each side and beyond the chin.

The ears—the most striking peculiarity of the fancy breeds—must be scrutinized very closely, to see that they are perfect according to the fanciers' rules, which are, that they must never measure less than fourteen nor more than seventeen inches in length from tip to tip, measured across the head, except for the oar-lop variety, when the extreme length may be eighteen inches; in point of colour, the ears must always be like the very darkest tints of the fur on the body,—if darker, so much the better, and of course, perfectly free from white markings, for if any light spots break the beautiful tone of colour, they produce a piebald appearance, which is a great defect. Fanciers reckon three grades between the common up-eared rabbit and the flat or perfect lop; these varieties are the half-lop, the forward or horn-lop, and the oar-lop.

It being one of the most important points, with regard to the ears of the fancy rabbit, that they should correspond exactly with each other, not only in shape, but in direction, the HALF-LOP from its having one ear upright, and the other pointing downwards, is the least in estimation, even if it is well-shaped and beautifully marked. However, as animals of this variety are generally very well-bred, if their colours and shape tally with the rules, they may be kept for breeding with advantage, as they not unfrequently throw first-rate lops.

A rabbit is termed HORN-LOPPED when its ears descend obliquely forward, from the side of the head, and project out far beyond the nose. This peculiar carriage of the ears is held in rather more estimation by some fanciers than the preceding, and it is worthy of notice, that almost all horn-lops occasionally raise one ear upright, and so resemble the half-lops.

The next style of carrying the ears is that of spreading them out horizontally on each side, and from the appearance they present when thus extended, the rabbit is termed an OAR-LOP; and such animals are considered very valuable, if correct in their different properties. Many of the best bred bucks are oar-lopped, and the same may be said of numbers of excellent does; for a good rabbit of the perfect lop kind, often carries one ear correctly, and elevates the other almost enough to entitle it to be classed amongst the oar-lops.

The FLAT or PERFECT LOP, is the most valuable of all the fancy strains, and as this carriage of the ears is exactly the reverse of the natural position, rabbits possessing this property, and perfect in every other respect, are highly prized. The ears of a first-rate lop must be so turned that the hollows of them are backwards, and the outer or convex part in front, and of course, correspond with each other in fall, and the closer they keep to the side of the check, so as to incline but little outwards, the more beautiful the animal is reckoned. Five, and ten guineas, and even more, have been paid for particularly fine specimens of this variety of the rabbit. If a fancier possesses ten or twelve does, all of them perfect, or nearly so, in their properties, he may consider himself fortunate if they produce him half-a-dozen first-rate lops in a season, for he must not expect that all the young ones in a litter will be thoroughly perfect; if he does, he will be disappointed, as it most generally happens that only one or two turn out of any value, the others being deficient either in colour or the position of the ears. Although it is perfectly impossible to insure the continuance of a fancy strain throughout all the litters, yet it is advisable to take the utmost possible precautions whereby so desirable an end may be in some measure secured. The bucks and does should therefore be of the best blood, and the does not allowed to kindle more than three or four times a year. The food should also be particularly attended to, and the most nourishing which can be procured given to them. Amongst the fancy rabbits we must not omit the French variety distinguished by having long hair, which curls almost like wool. By some persons this breed is supposed to be a cross between the beautiful rabbit of Angora, and the common white species.

RABBIT HUTCHES AND COTES.

Rabbit hutches should be made very neatly, not merely for elegance, but for the important reason of cleanliness; however, as it is not in every lad's power to obtain well finished ones, comfortable, though homely-looking hutches may be easily constructed out of egg-chests, which may be obtained at any cheesemonger's shop, or out of old tea chests; the former will serve as habitations for the does, and the atter for the bucks, or else weaned rabbits. The doe's hutch should be a foot and a half or two feet high, about two feet deep, and at least four feet long; about one-third of this length should be partitioned off to form a sleeping apartment, and in order that the animal may have free access to its dormitory, an aperture should be made in the partition, of just sufficient dimension to allow it to pass through with facility: a sliding panel or hanging door over this hole will be found extremely useful, as by such a contrivance the rabit may be confined in one division, while the other is undergoing a thorough cleansing. The edges of this aperture must be cased with tin, as rabbits are particularly fond of employing their teeth upon the woodwork of their prisons, nibbling every part they can lay hold of. The front of the hutch may be said to be composed of two doors, that is, a large door or framework of wood, having iron wires placed perpendicularly, about three-quarters of an inch asunder, is made to fit from one end of the front to the before-mentioned partition, and another door of wood without wires, from the partition to the other end. These doors must open in contrary directions, for which purpose the hinges of each door should be fastened at the ends of the hutch, and that one may be opened without the other becoming unfastened also, two buttons must be put on the partition, as shown in the illustration. The wired door should not be so deep as the wooden one, as a drawer for food will be required to slide underneath it. The edges of the feeding trough require to be cased with tin, for the reason before given; and as some rabbits scratch their food out of the trough, and soil in it, if the front edge of the trough, be bevelled off, and a piece of thin board an inch in width, and also cased with tin, fastened so as to lean over the top of it, it will be found a good safeguard against such habits. The floor of the hutch should be of smoothly-planed wood, and made to slope towards the back, along the whole length of which a narrow slit should be made to let the wet run off; a large smooth slate, however, from its being impervious to moisture, makes a far sweeter and better flooring than wood, besides which, it has an additional recommendation of being much more easily kept clean.

The buck's hutch, as most usually made, is about two feet and a half broad, one foot eight inches in height, and one foot eight in its deepest measure; in shape it differs considerably from the doe's, and may be easily understood by reference to the annexed figure. It is not divided off by a partition, neither is there a drawer for food running the length of the front, the receptacle being placed in the centre of a cross piece, which goes from side to side; the door, which composes the whole front of the hutch, must have very strong hinges, and the button to secure it be perfectly firm in its fastening; the wires should also be stout, and they may be set rather wider apart than in the door of the doe's hutch. The back of this hutch is nearly semi-

circular in its form, and an opening must be made at the lower part for the purpose before described. The hutches, if many rabbits are kept, may be piled one upon the other, but none of them ranged upon the ground; the lowest being placed on a stand about two feet in height, to keep out the rats and other marauders; and still further defences against such unwelcome visitants may be employed in the shape of circular shields of tin, about the size of a common plate, surrounding and affixed to the legs of the stand, as we have shown in our illustration, and which no rat can get over, or few things walk upon—excepting a fly, with its back downward. The backs of the hutches should not be put quite close to the wall, but a space be left between, that the dirt may pass, and be easily cleared away. If the rabbits are allowed the range of an outhouse or cote, it will be requisite to stop up all holes in the brickwork and flooring with little bits of brick, or tile, and then coat the flooring over with cement, to hinder the rats from getting in, and the rabbits from burrowing their way out. It is especially necessary that the rabbitry be thoroughly dry and well ventilated, not only when the doors and windows are open, but also when they are shut; the best proof as to the sufficiency of ventilation is the atmosphere of the house when you first enter it in the morning, and if any strong or unpleasant smell pervades it, you may be certain that a proper supply of fresh air does not circulate through the place, and an additional opening must therefore be made, which may be suffered to remain open by day and night; all such openings, windows, &c., to be protected by fine wire lattices, and so disposed that no draught blows directly through the place, for if the animals are exposed to chilling currents of air, the young fancier must not expect that they will thrive. The feeding trough should be heavy, and made of such clay as bricks are, for by being weighty it is not so likely to be overturned by the little captives when frolicking about. An artificial burrow for the doe to form her nest in should not be omitted in a rabbitry, but the arrangement of such a contrivance must be left to the ingenuity of the youthful proprietor, and the means at his disposal. If a little space of paled-in ground can be allowed in front of the rabbitry for its inmates to sport about in during fine weather, it will be of great advantage to them.

DISEASES.

Care in selecting the food, regularity in the hours of feeding, and attention to the general cleanliness of their habitations, will in a great degree preserve rabbits from disease. From the great value of the fancy rabbits, they deserve, when suffering from any malady, much attention and all the remedies which experience has proved to be the most efficacious; indeed, no boy would willingly allow a dumb creature to perish, if it lay in his power to preserve it, however common it might be.

HOARSENESS is a disorder which arises from the rabbits having fed too plentifully upon green food, and its symptom is that the animal's dung is moist and discharged too often. A liberal allowance of solid food, such as barley-meal, oatmeal, bran, &c., is the best remedy, with a sprig of parsley or fennel, and a small quantity of good sweet hay now and then; two tablespoonfuls of water, and not more, a day, may also be administered, as likewise oatmeal and green peas made into a stiffish paste.

For the LIVER COMPLAINT there is no remedy; the only thing which can be done is to promote the health of the animals as much as possible by keeping their houses and hutches warm, dry, and clean, for everything which adds to their general health acts as preservative against attacks of this disorder.

The SNUFFLES are occasioned by damp and cold, and whilst this complaint lasts the rabbits must be carefully dieted and secured from damp and variations of the atmosphere; boiled potatoes and bran, made into a paste, or barley-meal, or oatmeal, and ground peas mixed up into a stiff paste, with a little milk or water, will be found the best food, and no water nor green food of any kind should be given. As the animals recover, their diet may be changed by degrees, giving at first a little clover or sweet hay, and slices of carrot, and then gradually the vegetables to which they have generally been accustomed.

POT-BELLY is a disorder to which most young tame rabbits are subject; the symptoms are enlargement of the belly, and weak, and poor appearance of the sufferers. The most certain restoratives are air and exercise, which they should therefore be allowed whenever an opportunity offers. Much dry food, and a very small allowance of water, prove likewise very beneficial; and the only vegetables which can be given with safety are carrots and parsnips. Unless the poor animals are soon cured of this complaint they die.

RED-WATER is a complaint of the kidneys, frequently caused by wrong food, or damp and cold. The rabbit suffering from this malady must be put into a warm, dry, comfortable hutch, and supplied with oatmeal, bran, baked or boiled potatoes, &c.; two or three table-spoonfuls of water in which bran has been soaked, may be given every day; in the summer a few leaves of the milk thistle and lettuce, will be of great service.

LAWS RELATING TO RABBITS.

By the common law, if rabbits come on a man's ground and eat his corn or herbage, he may kill them. By the 7 and 8 George IV. c. 26, sec. 36, if any person wilfully and unlawfully, in the night time, take or kill any rabbit in a warren, or place kept for breeding rabbits, whether inclosed or not, he is guilty of a misdemeanour; and if in the day-time, the offender shall forfeit such sum, not exceeding five pounds, as to the justice before whom he may be convicted shall seem meet.

"We are indebted to this little insect for our greatest luxury in clothing, a reflection which ought to humble our pride; for how can we be vain of the silk that covers us, when we reflect to whom we are indebted for it, and how little we are instrumental in the formation of those beauties in our clothing of which we are vain."—STURM'S *Reflections.*

SILKWORMS.

BREEDING silkworms was quite a mania in England a century and a half ago; ladies of title and fashion had cocoons hanging about their apartments, and many a flunkey was as crimson in the face as the scarlet continuations below his waist through looking after the fires to see that a regular heat was kept up, and climbing the trees to gather mulberry leaves. One poor flunkey was tried because the worms left to his management perished, and there was a strong suspicion that he had given them poison, but whether with a pap-spoon or a soup-ladle was never clearly proved. A patent was granted by George I. for establishing an English silk manufactory, and two thousand mulberry trees were planted at Chelsea to feed the silkworms; but the manufactory and the trees came to nought, and all that came of the affair was a capital work written by Henry Barham, and published in 1719, on the management of silkworms. There was a society that gave silver and gold medals to those who grew the greatest number of mulberry trees, which at this period were propagated from cuttings. Are there any of the two thousand mulberry trees that were planted in the time of George I. standing at Chelsea now, we wonder? There was a Mrs. Williams, of Gravesend, who made a stir in the world of silkworms some eighty years ago, and who hatched the pretty dears so early that when they kicked up their little heels for the first time in this wicked world there wasn't a mouthful of anything for them to eat, as the mulberry trees were as bare as the bridge of your nose, so the little worms gave it up for a bad job, and as they couldn't creep back again into their eggs, why they died, and were done for. Then there was a Miss Rhodes, who had thirty thousand silkworms, and who calculated that they would produce her about five pounds of silk, and who was—we quote her very words—"determined not to relinquish her design until she had obtained the quantity of silk necessary for a dress;" but such a cold July set in about the time her worms were ready to spin that, like Cardinal Wolsey in Shakspere's "Henry the Eighth," she was ready to exclaim as she looked at her mulberry trees:—

> "To-day they put forth
> The tender leaves of hope; to-morrow, blossom,
> And bear their budding honours thick upon them;
> The third day comes a frost, a killing frost,
> And—when I think, 'Good mulberry trees,' full surely
> 'Their leaves are now all branching'—nips their roots,
> And then they fall as I do."

She tells us that "her distress increased hourly," and her thirty thousand decreased to some five thousand; but whether in after years, by hoarding up, she obtained silk enough for a dress from future generations of silkworms the world will never know, as she long since went the same road as her silkworms.

So far we have given you all we know about the silkworm in England. Something we might say about the Emperors of China and what they did for silkworms about the time when our Saxon king Edgar was making laws for the destruction of wolves in England, but we don't think you would be a bit wiser for the information, for it's all "hear-say" after all, and "hear-say," when above a thousand years old, isn't worth the parings of a midge's toe-nail.

Silkworms' eggs may be purchased very reasonably at many places, one of which is Covent-garden market; and after a stock is once laid in, they may be preserved till the following year, if care is taken to keep them in a dry drawer or box during the winter months. When first 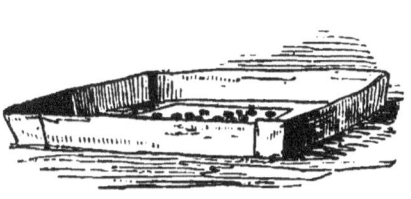 laid, the eggs are of a pale yellow tint, but they soon change to an ash colour. Towards the end of April, or early in May, just as it is a forward or a backward spring, when the mulberry-tree puts forth its leaves early or late, the eggs should be strewed or placed on the paper on which they were laid by the moth, in small and rather shallow trays, which ought to be made of good substantial cartridge-paper, with the edges turned up to about the height of an inch all round, as shown in the marginal illustration, and pasted neatly together at the corners. These trays containing the eggs should be placed in a window where the sun may shine full upon them, and if they can receive the rays of the mid-day sun, so much the better. Particular care must be taken to place them out of the reach of cats or birds; some persons, indeed, take the precaution of covering them with a piece of fine gauze, which, upon the whole, is not a bad plan.

Thus arranged, the eggs must be left until the hatching commences, at which time mulberry leaves, if ready, must be given the young worms, or if they cannot easily be procured, lettuce leaves must be given them; but they will not spin a silk worth looking at on this diet. Trays, made like those already described, must be obtained to receive the larvæ or worms as they come into life—a precaution necessary to prevent the unhatched eggs from being disturbed by the young worms. The operation of removing them must be performed very gently, by means of a feather or of a camel's-hair pencil, because the chrysalis at this early period of its existence is exceedingly delicate and tender. The hatching can either be hastened forward or kept back, so as to suit the leafing of the mulberry, by not disturbing the eggs until the leaves are nearly ready.

The first tint of this insect is darkish, which, however, turns afterwards to a creamy white; it has on each side, at every joint, a small circle, two half circles on its back, six feet (three on each side near the head), and ten holders (eight in the middle of the body and two near the tail).

The silkworm suffers four sicknesses from the first period of its existence to the time of beginning to spin, and during each of these, which generally continues about three days, it does not eat, becomes thicker and shorter, and casts its skin. If leaves are given to them once a day before the first sickness, it is sufficient; after it, until the third, they should be fed twice a day, increasing the quantity of food in proportion to the growth of the chrysalis; from the third to the fourth periods of sickness they must be supplied with leaves thrice a day, and if the weather is excessively warm, four times at the least; from the fourth crisis, until they commence their spinning labours, the food must be given very frequently, as they then consume more than in the whole previous time of their existence.

During the period of this moulting, as the change is called, the worms should be kept in a room where the temperature is not less than from ninety-five to one hundred degrees; the lower the heat under this, the longer they are undergoing their natural changes. Thirty-two days after hatching they will, if properly managed, have attained their full size.

Although we have said that lettuce leaves may be given to the silkworms during the first few days, yet as their natural food is mulberry leaves, they should be provided with it as soon as possible. It must be especially borne in mind, that they must not be fed upon lettuce *after* they have once been furnished with mulberry leaves, for such a change of food not only disorders, but destroys them. The trays ought to be cleaned out regularly every morning, until the last moulting of the chrysalis, when, as the dirt accumulates much quicker, they require greater attention; at which period also, they should be kept exposed to the air, particularly if the weather is at all favourable. In order to clean out the trays, the silkworms should be moved with the greatest tenderness; to effect which, when the insects have arrived at about one-third of their full growth, it is necessary to put new leaves into the trays, upon the tops of the half-devoured ones; the insects will soon crawl on to the fresh leaves, when they may be safely lifted out, and placed in their clean quarters in other trays. When they are full grown, they may be taken up in the fingers, caution being observed not to squeeze them, or let them drop. The leaves must if possible be fresh, but if there is no tree handy, and you have to keep a stock of leaves, let them be kept closely packed together, in a clean cloth.

When the insects are ready to commence their spinning occupations, they turn to a clear pink, or rather flesh-coloured hue, (particularly at their tails,) become exceedingly restless, and abstain from food. On this last symptom taking place, you must remove such of the worms as evince it into little paper bags made in the shape of funnels, wide and circular at the mouth, and terminating in a pointed end. 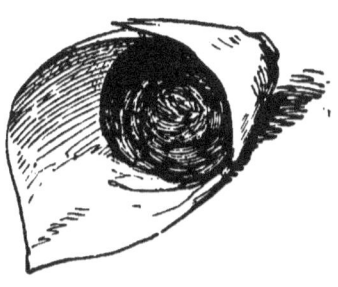 The depth of these little bags should be about four inches, and they are usually pinned to a tape, horizontally secured on the wall of a room. Here the little artists prepare a retreat, by disposing their silken threads in such a manner as to enclose themselves completely in an oval shaped ball of silk, of about the size of a pigeon's egg. This is is called the "cocoon." Within it the chrysalis once more casts its skin, turns thick, short, and of a dark brown, hard, glossy surface; becoming through this second change an aurelia. When the cocoon is about the size before mentioned, but not sooner, you may shake it gently, to ascertain whether the spinning  withinside is complete; and if a slight rattling sound can be heard, as though there were something loose in the cocoon, the insect's task is done.

The cocoon contains three varieties of silk, the one loose and unserviceable, the second closer and running crossways, the third very fine and gummed together, and this forms the inner coating. Care must be taken in winding off  this silk; the loose outer portion must be removed, then the cocoon placed in a basin of lukewarm water, that the end of the silk may be more easily detected, and the winding off facilitated, which may be done upon a common card; the length of the thread of a single cocoon varies from six hundred to one thousand feet. The aurelia, when taken out of the cocoon, should be placed in some bran, just under the sur-

I

face, where it will effect another change, and in about twenty days become a lumpish, inelegant, white moth. At this stage of its existence it does nothing more than deposit its eggs, if a female, and then dies. So soon as the moths emerge from the shell, they should be removed to the same species of trays as those in which they were previously kept, but are useless now, as the insect never eats leaves, the bottoms of the trays must be covered with clean white paper to receive the eggs; if, however, you wish to preserve a great quantity, it is better to place the moths in a coarse cloth, for the reason that by immersing the cloth in fresh water, you can destroy the viscous matter that glues the eggs to it, and then dry them well, and keep them together in a box as if they were beads, which is a more convenient plan than to be encumbered with a great number of papers or trays. If this method is adopted, the cloth must not be immersed until the eggs have assumed their ashy colour.

In some places, where silkworms are reared for commercial purposes, such a number of aurelias only are preserved as are necessary for the production of eggs; the others are destroyed by putting the cocoons in hot water, which process greatly accelerates the winding off of the silk; this plan must be adopted if you have a great number of cocoons, for unless you wind off the silk within ten days or a fortnight after you have ascertained that the worm has ceased spinning, the aurelia, even in its silken enclosure, will turn into a moth, and by piercing through the cocoon, destroy the silk. The silk varies greatly in colour, being of different tints, from white to a rich yellow, but the lighter colours are the most sought after.

The long and tedious process of winding off may be greatly abridged by means of a kind of reel sold at most Tonbridge-ware shops, which enables the operator to wind two or three threads of silk at the same time.

THE AQUARIUM.

THERE are few things that throw a clearer light upon animal and vegetable life than an aquarium, whether it be filled with sea water or fresh water. Men, in a confined space, where there is not a sufficiency of fresh air, die through suffocation; water-animals, whether covered with shells or scales, also soon perish in a limited supply of water, both man and fish being destroyed through the excess of carbonic acid gas which they throw out, and by which they are poisoned. But this never occurs—save through age, decay, or disease—while there is a sufficiency of oxygen to inhale, as what we draw in overpowers and renders harmless the poison we throw out, which can be proved at any moment by shutting ourselves up in a confined space where there is but little air, and remaining there until we find it difficult to breathe; then letting in a current of fresh air, when we once more breathe freely. We will confine ourselves to the word fishes, whether we speak of a shrimp or a shark, a whelk or a whale—though, by the way, a whale is not a fish at all, as it brings forth its young alive, and suckles them as a cow does a calf; but no matter, all our aquatic animals which are kept in an aquarium are understood to be fishes. Now, fishes can no more live long in an aquarium, in sea or fresh water only, than you could in an empty water-butt that was bunged, pitched, and made air tight; the carbonic gas you throw from your body would destroy you in no time, while if the bung was taken out and fresh air let in you might live there as long as your friends liked to keep you, though it would be a great relief at times to apply your mouth to the bung-hole and have a good mouthful of fresh air. But there are no end of sea-weeds—call them marine algæ if you like—that take in this poisonous carbonic acid gas, and give out immense quantities of oxygen, as you may see by the thousands of bright silvery bubbles with which these sea-weeds are covered, and which they are ever sending to the surface like fairy balloons, from their sea-green garden-grounds—the fishes the happy spectators. This is the life of the water, and alone enables the fishes to "live, and move, and have a being;" without their sea-weeds they would be as dead as the old Egyptians who built the pyramids. Sometimes this oxygen will be poured out in such quantities when the sun shines too powerfully on the aquarium as to lift the sea-weeds from their moorings, like so many blown bladders, when, all spangled with silver stars, they will float on the surface, until, ceasing to be acted upon so powerfully by the light, the weeds will again drop down into their sea-green beds; and these sea-weeds, that throw out so much oxygen, and form, so to speak, the blood of fishes, sometimes spread over hundreds of miles of ocean, like those found in the Atlantic, that cover such an immensity of space as to be called, where they most abound, the weedy sea, beginning in the Northern Atlantic, and forming an embankment above twelve hundred miles in length. What countless millions of strange creatures must harbour in this great sea-forest! That beginning in the east stretches westward to the Bahama islands. This was the weed which frightened the sailors who accompanied Columbus, and who, when they moved slowly through it day after day, were afraid that it would form firm land, and that they should never more be able to return through it. But these gigantic wonders of the deep have nothing to do with the sea-weeds necessary for the life of fishes kept in an aquarium, and we have only described this great gulf-weed, as it is sometimes called, to show that neither desert nor forest on the face of the earth extend further or spread wider than these unexplored ocean prairies. As yet we know but little of these great sea-gardens, the countless leagues of vegetation over which myriads of fishes swim or find harbour and safety from their pursuers in the network of these unmeasured mazes.

But before telling you about what sea-weeds to get to keep the water fresh and sweet and the fishes alive, we must first make you understand how your aquarium is to be made. And, first of all, a glass globe, such as gold fishes are kept in, makes a capital aquarium, with rock sand and weeds at the bottom, and filled up with sea-water. But then these globes not only distort the forms of the fishes, but are not large enough to hold a good variety, and to show you clearly what wonders are hourly worked within the sea. The sides and ends of an aquarium must be of glass, though some have slate for the end pieces, as well as the bottom of the vessel, the latter of which is always slate. But when the ends are slate you cannot well see what is going on there, though perhaps if one end only were of slate it would afford a little more shadow and shelter for the fishes. The glass or slate must fit perfectly true, so that the metal by which the pieces are held together, shall only be brought to bear on the outside of the tank, and nowhere to come in contact with the sea-water, for if it does, farewell to all your fishes, there will be the deuce to play, and the old gentleman who is said to keep the Great Fire Insurance Office below might as well turn in one of his jets of sulphur, melted pitchforks, and the scum of his earthquake boilers, for there would be such a compound of poisonous gases as would make even the old one sneeze if he got his nose over the aquarium. The salt water attacks the metal, and the metal poisons the salt water, while the fishes and sea-weeds perish in the conflict. Surely some wood may yet be found that will resist the action of salt water and last for years, though such wood has yet to be discovered. The sides ought to be made of plate glass, as it is free from those faults found in sheet glass; and Betty ought never to be allowed to approach too near the aquarium when she is in a passion and has got the poker in her hand. The vessel ought not to be too wide if you wish to see the fishes perfectly, a foot or fifteen inches is wide enough for a small aquarium which is three or four feet in length; the depth is not so particular, though the tank must never be very shallow, as upon the slate that forms the floor there ought to be a pretty thickish bottom of sand, pieces of rock to which the sea-weeds cling, and grit or pebbles, as much like the bottom of the sea as possible, for the little inhabitants to burrow and hide themselves in when not disposed to "take their walks abroad," for it is a funny sight, I can tell you, to see a couple of shrimps meet when they are walking out, and to notice how polite they are to one another, bowing and scraping, and though you can't hear a word, no doubt saying, "How do you do this morning, I hope I see you quite well." Nothing saving sea-weeds and salt water, with the bottom prepared as described, ought to be placed in an aquarium until the expiration of ten days, or a week at least, if even it is formed of a glass globe. The water is not in a fit state to receive the fishes before that time; and it is better to keep baling out a few tumblers of water every now and then, and letting it fall in again from a good height, just as you give a head to a glass of ale, as this lets in plenty of air, and sets the oxygen at work. Put fish into sea water that is not properly prepared, and they will "kick the bucket" before twenty-four hours are over. As for a new aquarium, it takes nearly three weeks for the cement to dry, and then it ought to stand another three weeks filled with fresh water, and the water changed a good many times during that period, for the slightest portion of obnoxious matter will kill the fishes; and nasty as you may think sea water tastes, you will soon find, when you set up an aquarium, that it is no easy matter to get it fresh enough for fishes to live in, and that the sea is a great deal sweeter than most people think. No doubt in time glass vessels will be made big enough for any moderate-sized aquarium, and all this poisonous metal and cement be done away with.

The next thing is to put in your plants and sea-weeds, and if you have been fortunate enough to bring with them a portion of the rock, or stone, or whatever it may be, on which they grew, and to which they still adhere, you will have no trouble at all to get them fixed, for they have but to be placed in the tank where you wish them to be, and they will soon begin to throw out oxygen as fast as a glass of fresh champagne sends up beaded bubbles to the brim. Once tear them from their moorings—for roots, properly speaking, they have none—and although they will continue to live while floating in the tank, yet by no means can you ever get them to affix themselves to anything again, for many have tried and not one has succeeded hitherto; and this is why you should be so particular in gathering your plants, and have a chisel and hammer, so as to cut off and bring with them a portion of whatever they adhere to. Among the plants or sea-weeds so essential in keeping the fishes alive in the aquarium, we shall point out a few of the most beautiful, so that you may have something both useful and ornamental at the same time, for there are plants growing in the sea as beautiful as any that ever expanded their green leaves to the sun in the gardens of earth; and if there are blue-eyed sea-nymphs who sit combing their golden hair in coral caverns, and singing wild sea-songs, they must have lovelier gardens to float in than ever delighted the eyes of the sons and daughters of men. The Purple Laver is a splendid weed of a rich purple colour, and not looking much unlike the acanthus in form, which gives so rich a finish to architectural decoration. Of course you all know the classical story of the origin of the acanthus; if you don't, inquire—it is too long to be told here. The Common Green Laver, also known by the name of the Sea Lettuce, and about as common as the dandelion in our way-side walks, may be picked up anywhere on the stones and rocks which have been covered by the sea only for a few hours every day. It is a beautiful, delicate, green-coloured plant, and as full of tuckers and puckers as our dear old grandmother's night-caps used to be after Betty had crimped them over the Italian-iron. Once seen, and you would know it again by the feel only, even in the dark. There is no better plant in the whole depths of the unbounded sea, we believe, for a marine aquarium than this—it throws off bubbles of oxygen by thousands: so much so at times, that the little globules fairly lift it to the top of the water, just as you might be carried up in a balloon, which "all the king's horses and all the king's men" could not keep down if they tried their hearts out, when once it had made up its mind to go up with you. This pretty plant throws out so much oxygen, that we shouldn't wonder if it tried, at its throwing all the fish out of the tank bang through the ceiling, slap through the roof, carrying off tiles and everything, and sending them near enough the sun to be fried nice and brown and ready to eat when they come down again. Now, if that isn't a "whopping" fib, tell a bigger if you can, and if the wager was a turnip to a leg of mutton, we should win. Then there is the common Sea Grass, which is almost as plentiful as the grass in our fields. You will hardly find a little pool which the sea has left but what is fringed with it. Sometimes it is as slender as a thread, then broad as narrow braiding, and you have only to pull up a few handfuls of this sea-grass, put it into a basket, bring it home and shove it into your aquarium, and you will find such a family of lively little creatures that were hidden among the fronds, as will amuse you for weeks after in watching their antics. But the most beautiful of all the algæ is the Scarlet or Crimson Delesseria. There is no leaf hanging from tree or flower more exquisite in form than that of this hand-some seaweed. It has also a mid-rib, and spanning fibres like the oak-leaf, and hardly one person out of a thousand would believe it was a seaweed until they smelt of it. About midsummer it may be found almost everywhere by the sea-side in perfection, and it is not only useful, but as ornamental to an aquarium as a rose-tree in full bloom is to a garden.

The Common Bladder-wrack, which every boy knows who has been to the sea-side, through having stamped on the bladders or air-vessels, that rise like blisters upon the weeds, and sound like crackers; and the Oar-weed, which sometimes grows to an immense length, and is used for making knife-handles, must *not* be placed in an aquarium, as they would soon spoil the water, and a good-sized tank of sea-water costs money, especially when it has to be carried far inland.

As to the marine animals to be placed in the aquarium, the varieties to select from are so great, that the choice must be left to yourselves. You ought, however, to have the pretty little Goby, a few small flat fish, the Pipe fish, and the Sea-Stickleback. But of all things you must get a few Hermit-crabs and Spider-crabs; also Shrimps and Prawns, who make themselves quite at home in the tank; and, if you wish to see your watery subjects clearly, do not forget the Periwinkles, as they eat off the green matter that adheres to the glass and prevents you from seeing what is going on inside. The most interesting objects you procure will be the Sea Anemones, which at one time (when closed) look like dirty mushrooms, and at another (when they open) like a basket filled with gorgeous jewels. But above all things, don't forget the Sea Mouse, for though he is so fond of burying himself in the sand and gravel at the bottom of the aquarium, yet, when he does show himself, and you get him fairly between your eye and the sunlight, you'll be almost startled by the rich array of colours he displays. The plumage of the humming-bird is not so gaudy as the edging of hairs which rise along the body of this little beauty. To paint the colours would be impossible, as they change every time these ridgy bristles are moved—purple and gold, crimson and green, flashes of amber, and shootings as of silver threads, darting and flashing to and fro, until the eye at last fairly aches through gazing at so much splendour. The sea mouse is very fond of hiding himself, if not in the sand and gravel, under the bits of rock, or among the seaweeds; but having once found him, you can generally depend upon his being "at home," or somewhere about the same spot, for a whole week to come. A keen look-out must be kept for the remains of the little animals that die in the aquarium, and they must be removed at once, or the whole stock of fishes, or whatever they may be, will speedily perish—as, in spite of seaweeds, the water will become too foul for them to live in it; and to neglect for a single day the removal of what is dead may destroy every living thing in the aquarium.

Hitherto we have written about what can only be accomplished with great care and anxiety, and at considerable expense; we will now turn our reader's attention to an aquarium that costs next to nothing, and which any boy may stock out of the first stream he comes to. Get a glass globe, which costs but some eighteenpence or two shillings, even for a very large one; fill it with fresh water, and put in it any of the small water weeds which you see growing in the water; you can scarcely do wrong, for whatever plants you find in it will throw out a sufficient supply of oxygen to keep your fishes alive; and, unlike a marine aquarium, they may safely be plunged into the globe at once—whether they be fresh-water spiders, shell fish, newts, or fishes—without even so much preparation as saying "by your leave." Little eels look very pretty in a fresh-water aquarium, and nothing can exceed the beauty and grace of their motions. Minnows, bleak, small carp, gudgeons, loach, will live for months in these globes, if the water is attended to; so will gold and silver fishes. But beware of the sticklebacks, they are such fellows to fight; and if they can find no other fish to quarrel with, will fight one another—for a brother thinks no more about pitching into a sister stickleback than a spider does about giving "pepper" to a fly that comes blundering and buzzing headforemost into his web. No doubt numbers of gold and silver fish die in their glass prisons through a want of water-plants in their globes to throw out oxygen, and that this is the sole reason why they thrive so much better in ponds. Remember, without plants to absorb the poisonous gas thrown off by fishes, and to supply air to the water, nothing can live in an aquarium in health for even the brief space of twenty-four hours.

GARDENING.

*"When Adam delved and Eve span,
Who was then the gentleman."—Old Rhyme.*

A PUZZLING question was often asked in our boyish days, which was, "How the first hammer was made?" and if gardening was man's first occupation on earth, some curious boy will no doubt want to know, "Where he got his spade?" To this we can only answer that, perhaps, it was a wooden one. But where did he get his knife, or whatever it was, to cut it into shape? inquires some young Master Inquisitive. To this we can give a clear, plain, satisfactory answer, "We don't know."

Gardening was no doubt the first thing man turned his hand to; it is the first thing a child tries its hand at, even now; for only put it on the ground, before it can run even, and it begins to scratch the dirt up, having first scratched its nurse, with its pretty little fingers. If it cannot get out of doors it creeps to the grate, and commences gardening amongst the cinders. And what does it set there? Well, when it burns itself, it sets up a good cry.

Now we are not going to write about the Garden of Eden, never having seen it; neither are we prepared to show how Noah grew mustard and cress in the ark, which no doubt the monkeys who ran about loose often scratched up, because we don't know that he cared about a salad; but what we are going to say is, first get a piece of ground and next a spade. Now, we dare say any boy thinks he can dig, when he can do nothing of the kind; he would no more know how to open his first trench than we know which was the first fish that swam in the sea. We don't call digging putting your spade in the ground, pulling it out again, and turning the dirt over, as your sister would a cake in the oven. That's only making a hole in the ground and filling it up again. Now, we'll show you how to dig properly. If you've a wheel-barrow all the better; it will save you carrying the first few spadefuls of earth to the other end of the bed. How long's the room you are sitting in?—Fourteen feet. Very well; now begin to dig at which end you like, and remember, to do any good, you must dig what the gardeners call "two spit deep," that is, two spades deep, for we suppose you are going to grow something you can eat. Now, having dug "two spit deep," you have made a precious deep hole in the ground, and what you have taken out must be wheeled or carried to the other end of the bed, or we'll say the other end of the room, and left there until you want it, and that will not be until you have dug the whole bed over; then you'll find a deep hole at the other end of the bed, and nothing to fill it up with but the earth you have carried there, which will fill it up to a T. Now you begin to work fairly.

by throwing the first spadeful of earth into the bottom of the hole you have dug; this done, you throw the second spadeful, which you have dug up nearly a foot lower than the first, on the top of the other, and by doing so you bury the first spadeful as you would a dead puppy, under the second. By this means you have brought new earth, which has not been exhausted by having anything grown on it for at least a year, on the surface, which earth has double the growing strength in it to that which you have buried. The digging done, as we have directed, and the last trench filled up with the earth moved to the other end of the bed for that purpose, now make the bed a little decent, by raking off the stones and such like, then cast your seed in—radish, carrot, parsnip, or onion— as evenly and regularly as you can. And now the seed is in, can you get on the bed and do "The Gardener's Dance?" Never saw it! Well, then, we'll teach you the steps. Get on the bed, put your feet close together, so that the insides of your shoes touch each other, now keep them close together all the time, then move first one foot and then the other, about two inches in advance at a time, still making the sides of your shoes touch all the time. Now cut away as hard as you like, and in as straight a line as you can; up and down, off we go, and that is the "Gardener's Dance;" the seeds are well trampled in. Now take the rake and smooth down al the pretty little ridges you made through dancing, and the work is done.

Unless the ground is wretchedly poor indeed, a bed prepared in the way we have shown will grow almost every kind of vegetable, such as peas, greens, potatoes, though the two latter must be put in with a dibber, supposing the greens to be transplanted. The best dibber is a broken spade handle, cut down to a point, it makes a capital hole in the ground, into which you may either drop a broad bean, a potato cutting, or thrust in a cabbage-plant.

Any boy can lay a bed out straight, who has a piece of string, two wooden pegs, and half an eye in his head; it is not half so difficult as drawing out the diagram before playing at "Hop-skotch." As to hoeing and raking, it is an insult against common-sense to tell any lad how to do that, for little fellows who are not breeched will get the old gardener's rake and hoe the very instant his back is turned, and set to work as if "to the manner born." And pleasant it is after "the winter is over and gone," to see the primroses which have been lying in the garden all the year long, once more shoot out their fresh green crisp leaves; and the hyacinths heave up their concave dark green spear shapes and sheaths; while the crocus sends from its bulb the long grassy-looking shoots which are streaked down the centre like a ribbon, and the lupina, whose leaves are the most beautiful of all, darts up from its old root those twelve-leaved cup-shaped stems, which hold the spring-rain, and look like emerald vases filled with crystals. All these are up and out looking at the sun by the middle of February, if the weather is mild, while the sharp-pointed gladioli pierce through the ground like a spear, and the dusky southernwood shows its grey green amongst the earliest liveries of spring. All these are pleasant objects to look upon, after the black frosts and dingy snow of winter have left the ground crumbling and soft, while those old gardeners the worms come out, and with their castings throw up the richest and lightest soil in the garden.

Then, when March comes in, what a number of things there are to attend to! balsams to sow in a little hot manure, and, if possible, to cover with a hand-glass; verbena cuttings to be taken off and potted; geraniums to be dealt with in the same way, using loam and silver-sand to strike them in, and keeping them warm under a little frame;—and what boy cannot get a bit of glass or buy an old window-sash, put a few boards round for it to rest upon in a sloping direction, and facing the south?—then place cuttings and seeds in pots under the glass to be ready to plant out in the garden beds about the end of May. It is wonderful how forward seeds will be, placed in pots, under a square yard of glass, and what strong plants they make when the time comes for taking them out of the pots and planting them in the garden beds.

Then how pleasant it is to see the rose-trees and the lilacs in early March covered with young green leaves; the almonds and japonicas in blossom; the birds singing around you while you are digging, sowing, hoeing, or raking; the murmur of some solitary bee who has come out to see how nature is progressing, and to report to his brother bees what flowers are beginning to bloom; while a butterfly or two, that has been hidden no one knows where, darts round you as you set the sweet-peas, and perhaps thinks what a jolly swing he will have on them when the painted ladies are in flower. The gooseberries and currant-trees are also covered with green buds, bringing before the eye dim visions of pies and puddings, rolly-pollys, and great fat-looking pots of jams, which it is so pleasant to get into a corner with, and, having had a shave from the new loaf, finish the lot, even to licking the spoon on both sides. Heigho! pots of black-currant and raspberry jam are wicked temptations to place in the way of hungry, growing boys; and those who leave them so easily to be got at are not free from blame.

Then, it seems hard not to have enough of what can be grown without any trouble at all. Take and cut a straight young shoot, about a foot long, from off either a gooseberry or currant-tree in winter, cut off three or four buds at the thick end of it, just long enough to go into the ground, whip off an inch of the top or thin end, and wherever you can find earth enough to stick it in, there it will grow and become a tree, and in a couple of years bear fruit. Vines and roses can be propagated in the same way; it is only cutting a stick off (a young one), cutting off three or four buds at the thick end, sticking it in the ground, and there you are. These are called "cuttings," and the last year's shoots hardly ever fail. Geraniums and verbenas may be cut off and grown in the same way, though a little silver-sand is necessary to get them to strike; and when once the "cuttings" have sent out into the sand a few white strong fibres, they may be put into little pots at once, in garden mould and loam, and when the roots have filled these small pots they must be removed, and either "bedded out"—that is, put into the garden bed, or be placed in larger pots.

Layers are managed in a similar manner, with this difference, they must *not* be cut clean off. We hardly know what plant may not be grown by layers, neither does it matter whether you make your cut high up or low down, though there is no doubt the sap sooner reaches the incision made in the layer, and causes it to strike earlier, the nearer it is to the root. To make a layer, pick out a young stem, cut it lengthwise nearly half through, drawing the knife in the direction towards the top of the stem, this sloping half-cut had better not exceed an inch in length; having made this incision, put anything in to keep the cut open—a peg, a bit of stone, no matter what—then all you have to do is to bury the cut portion in the earth, and as the other end of the layer is still a portion of the original plant, being attached to it by the root, and the top or thin end above the cut is out of the ground, the new root will be formed where the cut is made, and in time may be separated from the original layer that is connected with the root or stock of the plant. In brief, the art of layering is pegging a stem down in the middle and making a root, where had it been left standing there would only have been a branch. If you notice the strawberry, it throws out runners, at the end of which, or very near, you will see an offset, that sends out a root and settles in the earth of its own accord. Layering accomplishes the same end; the branch we peg down is the runner, so to speak, the incision we make, and set in the ground, is the offset, and forms the root of the new plant. Nature does for the strawberry what we do by art to no end of plants; it sends out a layer or runner from the strawberry-root, which runner sends out a root and plants itself, without any trouble to gardeners or owners. To conclude, layers are only cut half through in a slanting direction when the part cut is buried in the earth and pegged down.

Pruning, grafting, and such like, are beyond the management of boys, and to understand these operations thoroughly a good work on gardening must be studied. As to setting seeds, almost anything may be put in the ground in March and April, and will be sure to come up if the beds are properly prepared, the seeds left undisturbed, and not drowned by too much watering. For our part, we never water anything until we see them beginning to droop for want of nourishment, then we give them a thorough soaking, and that lasts a long time. Remember, if you want fine flowers, to pull off all your plants that have bloomed and are decaying; you will have no seed by doing this, but a succession of splendid flowers, and as fine as the first.

www.ingramcontent.com/pod-product-compliance
Lightning Source LLC
Chambersburg PA
CBHW021939160426
43195CB000118/1150